QUANTITATIVE

ITATIVE

GEO

GRA

PHY

In Association with **The Spatial Analytics and GIS series**

In an age of big data, administrative data, volunteered information and complex surveys, spatial analytics is a growing field of interdisciplinary research. Its focus is to make sense of and extract knowledge from spatial data through an expanding toolkit of analytical, computational and visual techniques. **The Spatial Analytics and GIS series** provides a user-friendly introduction to these techniques, in a way that is accessible to students in the geosciences, geographical information science and quantitative social science. Written by world-leading experts, and balancing theory with application, the series provides a 'go to' resource for people who are interested in new data and new methods for analysing spatial processes and relationships.

QUANT ITATIVE

GEO GRA PHY

RICHARD HARRIS

THE BASICS

Los Angeles | London | New Delhi
Singapore | Washington DC | Melbourne

Los Angeles | London | New Delhi
Singapore | Washington DC | Melbourne

SAGE Publications Ltd
1 Oliver's Yard
55 City Road
London EC1Y 1SP

SAGE Publications Inc.
2455 Teller Road
Thousand Oaks, California 91320

SAGE Publications India Pvt Ltd
B 1/I 1 Mohan Cooperative Industrial Area
Mathura Road
New Delhi 110 044

SAGE Publications Asia-Pacific Pte Ltd
3 Church Street
#10-04 Samsung Hub
Singapore 049483

Editor: Robert Rojek
Editorial assistant: Matt Oldfield
Production editor: Katherine Haw
Copyeditor: Kate Campbell
Proofreader: David Hemsley
Indexer: Cathryn Pritchard
Marketing manager: Sally Ransom
Cover design: Stephanie Guyaz
Typeset by: C&M Digitals (P) Ltd, Chennai, India
Printed and bound in Great Britain by Ashford
Colour Press Ltd., Gosport, Hampshire.

Library of Congress Control Number: 2015958574

British Library Cataloguing in Publication data

A catalogue record for this book is available from
the British Library

MIX
Paper from
responsible sources
FSC
www.fsc.org   FSC® C011748

ISBN 978-1-4462-9653-0
ISBN 978-1-4462-9654-7 (pbk)

At SAGE we take sustainability seriously. Most of our products are printed in the UK using FSC papers and boards.
When we print overseas we ensure sustainable papers are used as measured by the PREPS grading system.
We undertake an annual audit to monitor our sustainability.

To my family, Rabbits included.

CONTENTS

LIST OF FIGURES

LIST OF TABLES

ABOUT THE AUTHOR

Richard Harris is professor of quantitative social geography at the School of Geographical Sciences, University of Bristol. He is the lead author on two textbooks about quantitative methods in geography and related disciplines: *Statistics for Geography and Environmental Science* (Prentice Hall, 2011) and *Geodemographics, GIS and Neighbourhood Targeting* (Wiley, 2005). Richard's research interests include the geographies of education and the education of geographers. He is currently Director of Bristol Q-Step Centre, part of a multimillion-pound UK initiative to raise quantitative skills training amongst social science students, and has worked with both the Royal Geographical Society (with IBG) and Higher Education Academy to promote numeracy and to support the transition of students from schools to university.

The University of Bristol Q-Step Centre is one of 15 Centres across the UK working to promote a step-change in quantitative social science training for undergraduates. The Q-Step Centres are delivering specialist undergraduate programmes, including new courses, work placements, and pathways to postgraduate study.

Q-Step is funded by the Nuffield Foundation, the Economic and Social Research Council (ESRC) and the Higher Education Funding Council for England, and was developed as a strategic response to the shortage of quantitatively skilled social science graduates.

You can find out more at www.nuffieldfoundation.org/q-step.

Q-Step

A step-change in quantitative social science skills

Funded by the Nuffield Foundation, ESRC and HEFCE

PREFACE

The plan was simple. Basic, even. I'd hide myself away in Ireland and write as much of this book as possible in one long splurge. Easy. Or maybe not: I went on the trip but now, almost two years later, I am typing these words at the weekend and on the morning of my birthday. That's a clue to how overdue the manuscript is. So what went wrong?

I could, I suppose, blame my hosts. I'm not yet persuaded that Guinness tastes better in the Emerald Isle, but I do know it tastes better in good company. And they definitely were good company. But, no, I can't blame others for my tardiness, so there are no excuses there.

It might be that I lack focus. That's definitely true. Running every day for a year in support of the mental health charity, Mind, is a bit of a distraction but a healthy one, I think. Today is day 318 of that challenge.[1] So far I have run 3545 km. It takes time out from my day, but I do some of my best thinking on the move. Besides, I can outrun the bus into work.

What happened was a change of context. Specifically, a renewed interest in quantitative methods in geography and in the social sciences more generally, and a concerted effort to increase numeracy and statistical literacy amongst students. Given that environment, an overly simple introduction to quantitative methods would lack ambition. It would sell the reader short and fail to convey what is a vibrant area of research and activity within geography. Consequently, the book you are reading is longer than I had intended and covers a wider range of topics, from the basic to the more advanced. I hope it is more useful as a consequence and gives a better impression of what quantitative geography is about.

My first book thanked almost everyone I had ever met. There is probably a dedication to my neighbour's brother's younger son's best mate's gerbil in there. The second was a little more constrained and dedicated to Les Hepple, a professor of geography whom I regard to this day as a model scholar and outstanding teacher. For this book, I will simply thank my family (Siân, Rhys and Timothy), my colleagues (Ron, Kelvyn, Ed, Clive, David, Winnie, Malcolm and Sean) and all

---

[1]See https://www.facebook.com/2015Krichharris/

the members of the running community, especially those at Emersons Green Running Club and Pomphrey Hill Parkrun who keep me broadly sane when the black dog threatens to bark.

RH
South Gloucestershire
(aged 42, and feeling it)

PART 1
ABOUT QUANTITATIVE GEOGRAPHY

1

INTRODUCING QUANTITATIVE GEOGRAPHY

1.1 INTRODUCTION

In 2010, two economists, Carmen Reinhart and Kenneth Rogoff, published an influential paper that provided a rationale for the policies of austerity adopted by many governments following the banking crises and global recession of the late 2000s (Reinhart and Rogoff, 2010). Their paper shows how increasing government debt stifles economic growth, but only once debt exceeds 90 per cent of a nation's gross domestic product (GDP, a measure of the monetary value of a country's economic output). You can understand how it caused alarm. In 2007, government debt as a percentage of GDP was 65.2%, 24.9%, 36.3%, 64.2% and 43.7% in Germany, Ireland, Spain, France and the UK, respectively. Three years later it had risen sharply to 82.5%, 91.2%, 61.7%, 82.7% and 78.4%, respectively – as shown in Figure 1.1. In Greece the percentage had risen from 107.2% to 148.3% over the same period. By 2013, and for most European countries, debt as a percentage of GDP had risen even further: Germany is an exception (78.4%); Ireland (123.7%), Spain (93.9%), France (93.5%), UK (90.6%) and Greece (175.1%) are not. The data, freely available on the Eurostat (European statistics) website, provide strong evidence that the debt to GDP ratios were worsening.[1]

But what if Reinhart's and Rogoff's thesis is not correct? In 2013, a university student, Thomas Herndon, attempted to replicate the study's findings. Herndon used the same data and the same methods of analysis. What he didn't obtain were the same results. He checked them. His professor checked them too. Still the answers differed. So he contacted the authors, and eventually the problem emerged. There was an error of calculation in the spreadsheet used for the original paper. You can read all about this on the BBC News website under the headline 'Reinhart, Rogoff … and Herndon: The student who caught out the profs'.[2]

---

[1] http://epp.eurostat.ec.europa.eu/
[2] http://www.bbc.co.uk/news/magazine-22223190

Mistakes happen. What is important is to find and correct them. Replication is an important principle of research, and a driving force behind free and open data, and free and open software. It allows people to go back and check analyses, to explore new lines of investigation and to look for other interpretations of the information available. The student had the data. He had the analytical skills, too. And, to their credit, Reinhart and Rogoff were willing both to provide the spreadsheet and to acknowledge their error, whilst also defending their conclusions: 'We do not believe this regrettable slip affects in any significant way the central message of the paper or that in our subsequent work.'

The story does not end there. In 2014, three of my colleagues also looked at the data, this time approaching it from a geographical angle (Bell et al., 2014). They could do so because they had the data, they had the skills and they had the geographical curiosity to look at the differences between places. What they found was that the relationship between debt and growth varies across countries. They write that 'these findings clearly indicate that, with regards to the effect of debt on growth, an argument and implied policy prescription based on a "stylised fact" alone is much too simplistic and general to be particularly relevant from a policy perspective … as such, the importance of debt for budgetary policies should not be exaggerated; much depends on time and place' (p. 22).

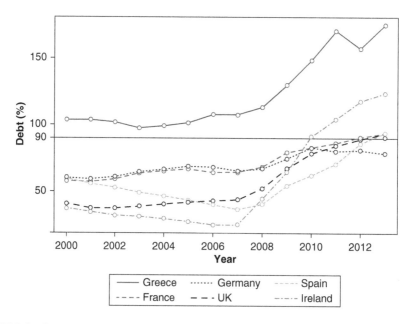

FIGURE 1.1 Government gross debt as a percentage of GDP for selected European countries, 2000-13. The 90 per cent threshold is said to be the one beyond which economic growth adversely is affected, but does that ignore important geographical differences? (source of data: http://epp.eurostat.ec.europa.eu/)

1.2 ABOUT THIS BOOK

This book has a simple objective. It is to provide an introduction to quantitative geography suitable for students beginning their undergraduate studies in the discipline. The aim is to get you thinking quantitatively; to help you understand and not take for granted some of the charts, numbers and statistical reasoning that appear not just in geography journals but in the media, in public policy, and in business and commerce. Data are collected for a reason, and even the most dispassionate researcher will have some idea, some question, some motivation for what they would like to get out of the data. As such, neither data nor analysis can be regarded as entirely neutral; results and research findings are always open to debate.

This is how it should be; research should be scrutinised and exposed to critical evaluation. It is how we sort the wheat from the chaff. In the run-up to the 2015 general election, Nigel Farage, leader of the UK Independence Party, was reported to have said, 'Here's a fact … there are 7000 diagnoses in this country every year of people that are HIV positive … 60% of them are not British nationals.' Yet, as an independent organisation shows, it is not a fact at all: the number of diagnoses was actually 6000 in 2013 (the most recent year for which there are data) and, of the 4980 for which birthplace was recorded, 54 per cent were born outside of the UK. Of those, the number who are not British nationals is unknown.[3]

This book encourages you to engage with quantitative geography through an understanding of it. Quantitative geography is the name given to a broad and multifaceted body of research that develops and applies quantitative methods to a wide range of areas of geographical enquiry. Quantitative methods are those that collect, analyse, present and help interpret data based on measurement and numbers. Such methods are widespread throughout the sciences and social sciences, and are used (albeit to a lesser degree) within the humanities too (for examples, see Gelman and Cortina, 2009). Quantitative geography can be found in social geography, economic geography, demography, environmental geography, hydrology, glaciology, geomorphology, transport geography, urban geography, rural geography, cultural geography, metrology, biogeography, climatology – in most parts of the discipline, in fact.

This book has the word 'basic' in its title, and it means precisely that: elementary, rudimentary, fundamental. The book does not pretend to be a comprehensive introduction to statistics, remote sensing, Geographic Information Systems (GIS), process modelling, computing, geostatistics, scientific visualisation, micro-simulation, geocomputation, econometrics or any other of the various fields and subdisciplines that intersect with and shape thinking in quantitative geography. It does include material

---

[3]See https://fullfact.org/factcheck/hiv_diagnosed_not_uk_nationals-41402, where there are links to the relevant data.

from those fields (notably statistics and geographical information science), but the goal is more foundational: to cover principles of numeracy, statistical literacy and computation, and to aid understanding of some of the key methods and concepts found in quantitative geography. The aim is not to replace or to compete with a textbook like Rogerson's classic *Statistical Methods for Geography: A Student's Guide* (2014), or my own co-authored text, *Statistics for Geography and Environmental Science* (Harris and Jarvis, 2011) but to provide grounding in some of the ideas and knowledge that permeate those books.

Another synonym for 'basic' is 'simple', but I am wary of describing this book in that way. It is dispiriting for a student if their instructor says, 'this is easy!' whilst the student experiences it as a struggle. Let me admit, therefore, that whilst none of the content is intended to be difficult, there could be parts that some readers find challenging, either conceptually or because of the mathematical notation involved. A recent survey of geography students found that those with a mathematical background tend to have more confidence when learning quantitative methods for their degree (Harris et al., 2013). For this group, familiarity with maths offers comfort with algebra, so it should be straightforward to read the basic mathematical principles found in Chapter 4. Others may experience anxiety about numbers, react unfavourably to equations, and be interested in quantitative geography only in so far as it is mandatory for a course in research methods.

To the second group I want to emphasise that whilst the mathematical ideas that sit behind the methods found in this book are undoubtedly useful, knowledge of them will not in itself help to draw a decent graph or facilitate the critical thinking that allows the nuance of statistical relationships to be explored. Moreover, I do not believe that those who got on less well with high school maths are somehow predestined to fail or falter when it comes to applying data to areas of geographical enquiry. Often what you learn at school is what is known as 'pure maths' – the laws and theories of maths, frequently taught in the abstract. Much of what you encounter in your geography degree will be more applied: it is about extracting knowledge from data. I shall not pretend that this book contains none of the language of maths. Perhaps it would be possible to write it with an absence of formulae and notation, but to do so seems incongruous, self-defeating even – a little like trying to learn French without ever using the French language (*plutôt imprudent*).[4] Nevertheless, this book is most definitely not a mathematics text (nor a statistical one especially, other than at an introductory level). It is more about numeracy and data literacy – learning how to use data to explore topics of relevance to geography, about how to draw conclusions from data, about how to communicate those findings, and about the use of software to support the

---

[4]That said, *Statistics without Mathematics* by Bartholomew (2016) is excellent. The sort of book I wish I could write.

processes of data collection, analysis and presentation. The mathematical demands are kept to a minimum.

1.3 WHY STUDY QUANTITATIVE GEOGRAPHY?

None of the analytical methods discussed in this book is original. Some of the statistical approaches have been around for more than a century. The maths is older still. A more recent development is the proliferation of data, much of it geo-tagged, so we know what is happening and where, but not necessarily why. It is hard to overstate this change: 'things' – including people – are measured and reported in ways, and with a frequency, that barely could have been imagined a decade or so ago, and often these data are available to be queried, analysed and presented in new and imaginative ways, tackling questions of fundamental interest to geography, as well as to other subjects. For example, 'smart cities' use measurement, monitoring and modelling to aid traffic flows, to provide real time information on public services, pollution and population movements, to manage bike sharing schemes, and to foster greater interaction between people and places. Making sense of the huge volumes of data these generate and what they tell us about how cities function has been described as urban analytics.[5]

Learning quantitative methods provides core knowledge and skills that are useful for geographical enquiry and for undertaking a research project. Many employers look for these skills in graduates. One reason why students attend university is to enhance their employment prospects – having said which, here is a statistical question: are historical data showing that past graduates typically earned more over their lifetime evidence that the same wage premium will continue today? It is true that having quantitative skills lends itself to gainful employment. And, in the UK at least, employers have signalled concern that too many graduates have insufficient quantitative skills for the needs of the job market. The British Academy (2013), a national academy of world-leading academics in the humanities and social sciences, put it like this: 'Generating and analysing data requires you to be numerate and statistically savvy. The skills that allow you to do this are highly prized in a range of careers.' That observation may provide a financial incentive to keep reading. However, let me avoid a misapprehension: I am not really writing this book to enhance your CV or to boost your pay packet. For me, learning about quantitative geography is about more than transferable skills and employability. Ultimately, I am writing this book because I want to share with you knowledge of a set of techniques and ideas that have shaped and continue to shape the intellectual terrain of geography, and are of increased relevance to the data-rich world we live in, and the opportunities for analysis and discovery that it brings.

---

[5]See http://bikes.oobrien.com for real-time visualisations of bike share schemes in cities across the world, including New York, Melbourne, Tel Aviv, London and Barcelona.

One confession, though: the book focuses throughout more on quantitative human geography than it does on physical geography. There are four reasons for this. First, the focus reflects my own interests and training. These will not be to the complete exclusion of physical geography; nevertheless, readers will detect a definite leaning towards topics of interest to the social sciences. Second, the reasons for introducing quantitative geography – easily stated in physical geography where the connections between numbers, data, analysis, science and geography are appreciated and rarely disputed – are arguably more pressing on the human side where quantitative geography has a long and rich pedigree but has also fallen in and out of fashion, reflecting wider trends and critiques of quantitative approaches across the social sciences. The cyclical nature of what is fashionable within a discipline leads to the third and fourth reasons to focus on the human side: a resurgence in quantitative social science in disciplines such as politics and sociology (as well as in economics, where quantitative research methods never went away) yet also concern that many students of the social sciences are ill-equipped to participate. Physical geographers may feel a little smug at this point, but analytical errors such as presenting data poorly, misusing statistics or going too far in drawing general trends from limited data are not difficult to find in this part of the discipline, leaving it exposed to these same concerns about the sufficiency of quantitative skills training.

The antipathy towards quantitative methods that persists in some areas of human geography is a reaction to success, specifically the period of what is described as 'spatial science', flourishing in the late 1960s and during the 1970s. Spatial science found inspiration in adapting scientific laws to socio-economic systems, drawing especially on economic theories to model, for example, urban systems. Some critiques of spatial science find it too tied to economics and, by implication, to the processes of power and wealth reproduction that (dependent upon your political point of view) sustain inequalities and social injustice.[6] Others find it narrowly focused on the measured and measurable, and on overly formalised methods of representation. It is criticised for taking a too narrow and depersonalised view of what it means to be human, eschewing sensitivity to individual experience in favour of more abstract conceptions like the principles of least effort and economic self-interest, and the actions of (supposedly) rational people.

There is much to be learned from these critiques. It is true that spatial science was commonly nomothetic – looking for general laws of behaviour – rather than emphasising the diversity of individuals' lived experiences. It also is true that spatial science tended to adopt the tenets of positivism within a hypothetico-deductive

---

[6]A recent briefing paper by the charity Oxfam (2016) said that in 2015 just 62 individuals had the same wealth as 3.6 billion people combined (half the world's population). This is a staggering but note how data, numbers and calculations are used to expose the inequality and to challenge it.

model of geographical enquiry – that is, of aspiring to the process of forming a hypothesis based on theory, expressing the hypothesis mathematically, attempting to test the hypothesis empirically using data and thereby accumulating evidence to deduce whether the hypothesis and the theory are valid or should be discounted. In general the approach assumes there is some sort of objective reality that can be measured, modelled, made sense of, predicted and possibly even controlled.

Whatever the merits or demerits of this type of thinking (which also is evident in disciplines such as economics and political science), and regardless of whether it was ever as widespread as textbook introductions to this period imply, it is an anachronism to characterise what takes place as contemporary quantitative geography only by the spatial science of its past. That is like reviewing the music chart from about 1972 and assuming it accurately describes what people are downloading today. (In 1972, the highest ranked single in the United States end-of-year chart was Roberta Flack's 'The First Time Ever I Saw Your Face'. I'd not heard it before but found it on YouTube. To be honest, it's not my kind of thing.)

Quantitative geography is a diverse field. There is no single philosophical perspective that binds its practitioners together. What there is is a sense that geography matters, and that geographical differences often matter. Much *geographical* analysis is about exploring differences between people and places, showing how modelled relationships are contextually dependent, and challenging the 'stylised facts' that ignore the variations. The paper by Bell et al. (2014) mentioned at the beginning of the chapter exemplifies this.

Figure 1.2 illustrates some of the components of quantitative geography: maths and numeracy; statistics and statistical modelling; data handling and (geo)computational skills; graphical presentation and the visualisation of data; geographical information science; and social and scientific knowledge. These are important and characteristically quantitative, but an extra element is needed before they can be considered as quantitative geography – that is, the ability to think geographically and to explore geographical lines of research.

Figure 1.3 gives an example of these parts coming together to produce maps of the average house price in London's neighbourhoods over the period 2001–13. To produce these maps isn't mathematically or statistically demanding. However, it does require gathering and collating the data, matching the location of house sales to neighbourhoods, classifying the neighbourhoods into groups by house prices, and mapping the results. Of interest is the spatial (geographical) variation that the maps reveal. There is spatial heterogeneity – the selling prices are different in different parts of the city. There also is spatial clustering – distinct enclaves of higher or lower house prices. It seems that the areas of most expensive housing have contracted into the centre of London. This is not because other places have become cheaper (house prices have risen there, too). It is because the most expensive areas have become *very* expensive compared to the rest.

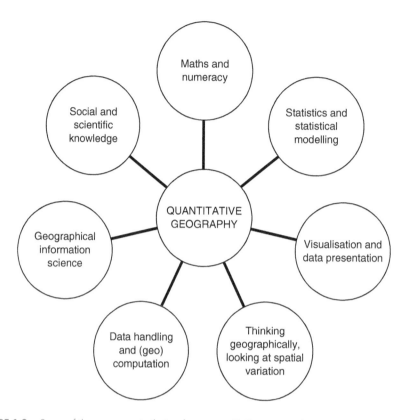

FIGURE 1.2   Some of the components that make up quantitative geography

The study of spatial variation is emphasised in the *Benchmark Statement for Geography*, a UK publication that outlines what is expected in the contents of a typical geography degree (Quality Assurance Agency, 2014).[7] It states that 'the concept of spatial variation is fundamental to the subject, meaning that geography graduates are able to demonstrate knowledge and explanations of spatial distributions in both physical and human phenomena' (paragraph 3.3). It goes on to stress the importance of numeracy:

It is important to emphasise numeracy and numeric skills. Geographers require skills in the presentation, interpretation, analysis and communication of quantitative data. They are familiar with a range of statistical techniques

[7]The Quality Assurance Agency for Higher Education is an independent body entrusted with monitoring, and advising on, standards and quality in UK higher education: see http://www.qaa.ac.uk/assuring-standards-and-quality/the-quality-code/subject-bench-mark-statements/honours-degree-subjects

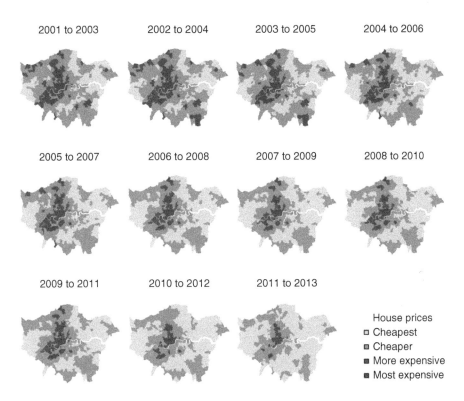

FIGURE 1.3 House prices by neighbourhood in London (source: author own calculations based on UK Land Registry data)

including simple descriptive statistics, inferential tests and relational statistics such as correlation and regression; principles of research design and ways to collect data; the retrieval and manipulation of secondary datasets; and geo-spatial technologies such as digital cartography, Geographic Information Systems (GIS) and remote sensing. Attention is given to spatial statistics, to issues of spatial dependency, to spatial difference and to the effects of scale (paragraph 3.13).

Here, then, is another reason for studying quantitative methods. To do so is part of what it means to do geography and to be a geographer. Having said that, I want to be clear. This book is not about positioning quantitative methods as better or more important than qualitative methods in geography. I don't believe they are. The *Benchmark Statement* also says that geography programmes 'should develop familiar-ity with a range of methodologies including those used in data collection and analysis; field and laboratory work and process modelling; programming; ethnogra-phy; focus groups; interviewing; archival work; discourse and textural analysis; and participant observation' (paragraph 3.12). I fully endorse that (in fact, I helped to write it). Moreover, there are other important knowledges, including the theoretical and the philosophical, that rightly demand space in a geography curriculum. In this book, I focus on quantitative methods not because I want to exclude other approaches but simply because quantitative methods are what this book is about. It is not intended as a textbook introduction to geography, just as an introduction to quantitative geography.

1.4 THE CONTENTS OF THIS BOOK

The book is broadly structured around the seven components of quantitative geography presented in Figure 1.2. Statistics are central to Chapters 2, 3, 5 and 6, with statistical modelling in Chapters 9–11. Maths and numeracy are considered in Chapter 4; visualisation and data presentation are in Chapter 7, and also Chapter 8, which is about GIS and geographical information science. Social and scientific ideas permeate all the chapters, as, hopefully, does thinking geographically. Chapters 8, 10 and 11 give particular focus to geographical forms of analysis. Chapter 12 discusses computation and programming.

The chapters build upon each other, but can also be read in groups if you want to dip in and out of particular topics. If you want a general introduction to statistics, why they are useful but also how they are misused, then finish reading this chapter and turn to Chapters 2 and 3. If you want to refresh your maths skills, read Chapter 4. If you want to cover the sorts of statistical ideas that you are likely to encounter in a standard methods course, read Chapters 3, 5 and 6. If you want to focus on data presentation and mapping, read Chapters 7 and 8. If you want a course in regression analysis, look at Chapters 9 and 10. If you want to focus on more explicitly geographical forms of analysis, turn to Chapter 11 (and also the second half of Chapter 10). For a very brief introduction to the programming language, R, see Chapter 12. If you read all the chapters, you will definitely have covered 'the basics' of quantitative geography and quite a lot more besides.

1.5 CONCLUSION

Back in Section 1.3, a question was raised but not answered. It asked, 'are historical data showing that past graduates typically earned more over their lifetime evidence that the same wage premium will continue today?'

It's an important question. In countries where university study was once paid for from the public purse and where fees are now charged to the individual, a justification for the change is that it is the individual who benefits financially from the education in the long term so it is right that he or she should pay. That principle

sounds fair enough, albeit that it represents a shift from viewing university education as a public good from which society as a whole gains to a private good that
benefits the individual. However, the justification is actually a predication, since
nobody can know whether recent graduates will attract a wage premium over their
working life until they actually do (or don't). The prediction is based on what happened in the past; the issue is whether it is a reliable guide to the future. It is hardly
surprising that graduates were able to take up well-remunerated posts when fewer
people went to university and the supply of graduates to the labour market was
smaller, but what is the effect on wages as the supply increases?

The good news is that graduates still appear to attract higher salaries: type 'graduate wage premium' into your favourite search engine and you will find plenty of
evidence that this is true. The bad news is the lack of guarantee – going to university does not necessarily result in a higher salary. You may also find evidence
that the premium is decreasing, that salaries have fallen for graduates or that it
depends on which subject you take, where you study and the level of degree you
attain. In America there is evidence that the average wage is rising but the take-
home pay of all groups, including graduates, is falling. In case you are wondering,
it's got nothing to do with taxation but something known as Simpson's paradox.
Chapter 9 provides details.

Quantitative research is important for informing or for evaluating policy decisions, and to help guide individual decision-making. However, the research is only
as good as the data. Predictions about complex and changing socio-economic
systems must contain an element of guesswork, no matter how well informed. The
same could be said of research into physical and environmental systems, including
models of climate. All data analysis contains uncertainty. Statistics can be understood as the process of drawing information from that uncertainty – of 'taming
chance', as one book describes it (Hacking, 1990). But we need to be wary. Not
all quantitative information is equally valid, and there is a real danger of using it
to create the sorts of 'stylised facts' that over-simplify to the point of absurdity or
sheer dishonesty. In reading this book I hope you will be equipped with the sorts
of tools that allow you to use quantitative methods within geography but also to
spot and be critical of where those same methods have been abused and misused.
If you'd like to read some good (or bad?) examples of the latter, then I'd highly
recommend Ben Goldacre's book, *I Think You'll Find It's a Bit More Complicated
Than That* (2014). He's right, it nearly always is, but the mistreatment of data is
made worse when journalists, politicians, scientists and geography students can't
even handle the basics.

2

THE USE AND ABUSE OF STATISTICS

2.1 INTRODUCTION

In a House of Commons committee meeting in 2012, the (then) British Education Minister, Michael Gove MP, was asked about his aspiration for all schools to be graded good or better on inspection. How is this possible, he was questioned, if the definition of a good school is one at which pupil performance exceeds the national average. 'By getting better all the time,' he answered.

Few would disagree that raising educational standards is a laudable aim. However, impossible targets are impossible to meet. By definition, not all schools can be above the average. Even if every one gets better, the average rises with them. It was an innocent mistake but one that caused the chair of the committee to quip, 'Were you better at literacy than numeracy, Secretary of State?'

Michael Gove is not alone in committing a statistical gaffe. In a 2013 speech, the (former) Mayor of London, Boris Johnson, raised a link between economic inequality and intelligence. 'Whatever you may think of the value of IQ tests,' he said, 'it is surely relevant to a conversation about equality that as many as 16 per cent of our species have an IQ below 85, while about 2 per cent have an IQ above 130.'

As it happens, it is of little relevance at all: IQ scores are standardised so that 16 per cent will *always* have an IQ below 85, and 2 per cent above 130, regardless of whether the true gap in people's abilities is increasing or decreasing. IQs are measured relative to the rest of the population, not by some fixed amount of intelligence (whatever that might be). Even if we all are geniuses, there will still be differences between us.

These and other examples of statistical misunderstanding are found in an academic paper arguing for better understanding of quantitative geography and how it has evolved (Johnston et al., 2014b). The instances are unfortunate but probably not worth getting vexed about, in much the same way we can excuse a headline that says something has increased by 200 per cent when, in fact, the number has doubled from $100 to $200. That's an increase of 100 per cent, not 200; what they mean is that the number is now 200 per cent of its original value. Yet even if these mistakes are forgivable, they should not be encouraged – it matters if you misreport the rise in a company's profits; it matters if you don't understand the numbers but use them to imply the poor are unintelligent;[1] it matters if your analysis is in error and if the conclusions you draw are inaccurate, partial or just plain wrong.

Simple mistakes can have costly consequences. An example occurred in 1999 when NASA lost its $125-million *Mars Climate Orbiter* at too low an altitude around the red planet. The lack of height was due to a conversion error, the failure to recognise that whereas one team was working in metric units, the other was using the Imperial system. An expensive error, but without loss of life.

Sadly, the same is not true of the *Challenger* Space Shuttle, which disintegrated shortly after lift-off in 1986, killing the crew of seven. The disaster occurred because of a seal failure on the booster rockets. The question whether to launch was discussed the evening before, because design flaws in the seals were known and the forecast temperature was unusually low. However, data from previous launches showed that when damage occurred to the seals, it did so across a range of temperatures, suggesting that temperature, in itself, was not a cause. Unfortunately, as the statistician David Hand (2008) notes, this is a case of drawing the wrong conclusion from incomplete data. First, the shuttle had never launched on such a cold day, so existing information was not necessarily reliable. More than that, the data did not include the number of times the shuttle launched without damage to the seals. The damage-free launches tended to be on warmer days. Whilst NASA's data showed damage could occur at various temperatures, that is not what they needed to know. The more pertinent issue was that it was more likely to occur at colder temperatures.

2.2 150 PER CENT OF STATISTICS ARE MADE UP

The sceptic will argue that you can prove anything with statistics. In practice, that's not true. In fact, it is very hard to prove anything with statistics. Reasons include the limitations of measurement, the problems of using data that are biased or incomplete, and because there might be multiple causes or explanations for the observed outcome.

---

[1]An even more controversial argument is that there are racial differences in intelligence that are due to genetics as well as the (social) environments in which a person lives. Such an argument was made by Herrnstein and Murray in their book *The Bell Curve* (1996), and criticized in books such as *The Bell Curve Debate* (Jacoby and Glauberman, 1995) and the second edition of Stephen Jay Gould's *The Mismeasure of Man* (1996), which is critical of the idea that intelligence can be measured as a quantity such as IQ.

Chapter 1 argued that a good reason to study quantitative methods is to be savvy about how numerical information is used and misused, and about the appropriateness or otherwise of the conclusions people draw from it. You are unlikely to agree that 'Whole Lotta Love' by Led Zeppelin must necessarily contain the greatest guitar riff ever, just because a recent survey said so. However, in other circumstances, numbers, data and statistics assume an air of authority that is less frequently challenged. This is unfortunate because, as the author of the paper 'Why most published research findings are false' (Ioannidis, 2005) points out, whilst many scientific studies are really too small to draw firm conclusions, the pressure for researchers to discover something new means 'non-results' are not published nor follow-up studies conducted, which means the results that do get published may not be typical (they could be the exception rather than the norm).

Statistical 'proof' is not like mathematical proof. It *is* possible to prove that the area of a triangle is $0.5 \times$ its height $\times$ length of its base. The formula is a necessary consequence of how the shape, height and base of a triangle are defined. Having established some starting principles (axioms), the formula follows. Its derivation is logical. Data can show it is correct, but the proof is independent of the data. In contrast, data are at the heart of quantitative geography. We use them to look for patterns, to test ideas, to make connections and to draw conclusions that are supported by the data but might be inaccurate or incomplete. We use data to seek out and test ideas that are presented to others to debate, develop or discard.

What statistics provide is not complete proof but evidence, evidence upon which to make judgements and to form knowledge. Carefully collected evidence strengthens an argument, but rarely settles it. This is because the way we view and understand the world is always framed in some way – by culture, by upbringing, by prior learning, by scientific consensus and by our cognitive limitations as a species. It is also because data arise from a decision to measure some things but not others, and that decision reflects what is of interest to a company, to science, to stakeholders, to policy-makers, to politicians, to funders, to the media and/or to society at the time. Measurements have a purpose. It is the purpose that makes them useful but also partial.

How others or we ourselves rate the statistical evidence will be affected by our support for the purpose, the quality of the data, the convincingness of the argument, and the consequences that follow. For example, I tend to trust the science of human-induced climate change and its prediction that atmospheric temperatures will rise. I am conscious that, during the time I wrote this book, there was uncertainty about what appeared to be a 'pause' in global warming (which may have been due to oceanic absorption of carbon dioxide), although that seems to have ended: the year 2015 was the hottest since records began according to the World Meteorological Organization. I am aware that there are vested interests both for and against the idea that humans are causing global warming. However, frankly, I have more faith in the motivations of scientists than I do in those of large extractive corporations and their shareholders. Therefore, I tend to view the evidence favourably.

Generally, I also approve of models of flood risk on the basis that it is better to know and be prepared. I was less impressed when one map of flood potential identified my house as at risk but not a previous property, further down the road. That information could have affected my insurance premiums, but I knew from prima facie evidence that it was wrong: the 'not at risk' house had flooded on two previous occasions; the other was uphill and always likely to remain dry.

The point is this: statistics are based on data and used as evidence in science, in society, in the media, in policy research, in business, in commerce and in academic writing. That evidence is not beyond debate, nor should it be. It should be handled with knowledge and with care. Some cases in point follow.

2.3 SCHOOLS, LEAGUE TABLES AND RANKINGS

Consider league tables of school performance, published in some countries. They act to cast schools in competition with each other, ranking them against measures of their pupils' attainment. The information is said to assist parents in making educational choices for their children. It also holds the schools to account, providing incentive for improvement. The incentive is not just reputational, it also is financial where the funding of schools is linked to the number of pupils they attract. Maybe league tables do work to raise standards. Maybe parents should be entitled to the information (different countries disagree on this). Even so, league tables are contentious; there are issues with the data and what can be learned from them.[2]

To begin with, there are questions about what is measured and why. In England, a measure was added, called the English Baccalaureate, to show the percentage of pupils completing a fairly traditional mix of subjects at high school (English, mathematics, history or geography, the sciences and a language). This was a deliberate political intervention to try and raise the numbers of pupils taking these subjects. There is nothing necessarily wrong with this intervention or in using statistics to try and force an outcome. It could be a good way of helping more pupils to gain places in universities or to gain the skills need for employment. The point is not to criticise, but to note that the measurement had a very deliberate purpose.

Second, there is a selection problem – we are not comparing like with like. The best-performing schools are those that have the best-performing pupils. That would be OK if we could be sure that the differences between schools were due only to their success or failure in getting the best from each pupil. However, the situation is more complicated than that. There is a link between wealth and the

---

[2]See http://www.bbc.co.uk/news/education-20628795 and Foley and Goldstein (2012).

educational advantages that wealth bestows upon children, including the ability of parents to afford the greater house prices around the most highly regarded schools. Housing markets create social geographies, the spatial differentiations of people based on what they can afford to own or rent. Consequently, the good performance of a school is a product of its locality and its attractiveness to the people who live there: it is a function of whom it teaches as well as how it teaches. It is for this reason that 'value-added' measures of performance are created, trying to separate out the genuine achievements of schools from what would be expected given their intakes.

Even so, the results and rankings are more uncertain than the league tables suggest. Finding differences between schools is of no surprise. It is improbable that each would score exactly the same. The question is whether the differences between the rankings have any substantive meaning? According to some experts the answer is no, not really: when uncertainties in the data are taken into consideration there is little to distinguish one school from another, other than at the extremes of the ranking.

Figure 2.1 is from a paper that ranks 250 schools, from lowest to highest, in accordance with how much value-added they gave their pupils (Leckie and Goldstein, 2011).[3] Importantly, the schools are representative of all English state-funded schools. That means the mix of schools considered is a good reflection of the mix found across the whole of the English system. Schools to the top right and above the horizontal zero line of Figure 2.1 do best. The average value-added per school is indicated by the circles, which, for the most part, look like a thick, black band running near the centre of the graph. There clearly are differences between schools; the values are not all the same. However, look also at the vertical lines extending above and below each circle. These can be interpreted as measures of uncertainty – of the amount of trust we can place in the ranking. They suggest a range of values around each circle that are plausible for each school. Note how these lines overlap. They warn us that the differences between schools and their rankings are not at all clear-cut.

Worse, the league tables can't tell a parent what they really want to know, which is how good the school will be for their child, not how good the school has been for others, collectively, in the past. Arguably, past performance is not much of a guide. As the authors observe: 'league tables have very little to offer as guides for school choice … even if several years of past school performances are considered' (Leckie and Goldstein, 2011: 836). However, others have argued that using performance tables is better than choosing at random, with pupils who choose the better-performing schools achieving higher attainments, on average, than those who do not (Allen and Burgess, 2013).

---

[3]The definition of value-added has since changed but that does not affect the argument.

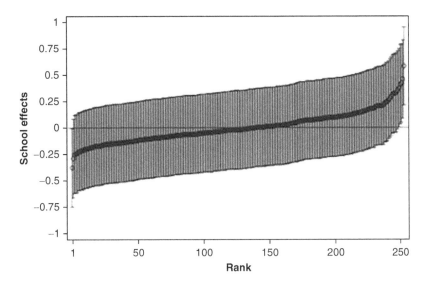

FIGURE 2.1 Showing differences between schools on a league table measure of 'value-added' (source: Leckie and Goldstein, 2011: 835)

2.4 LEAGUE TABLES, TRAFFIC CAMERAS AND REGRESSION TO THE MEAN

Imagine that a school comes bottom in a league table. The governors get together with the school authority, action is taken, and the following year the school rises higher in the rankings. Is this evidence that the intervention worked?

Before you answer, consider a different question. Imagine a person plays a game, tossing a coin ten times. Suppose the rules are that she scores a point every time the coin lands head side up, but, as it happens, it turns up tails each time. Her score is therefore zero from ten. She rests, has a glass of water, tries again and this time the coin lands head-side up for half the tosses. Five out of ten is a great improvement from zero. Was it due to her drinking the water?

Of course not. The player did better because she did badly before. She would be really unlucky to score zero again – that would be 20 tails in a row and an outcome far less likely than 10 tails followed by one or more heads. All that happened is her luck improved. Now think back to the school and apply the same logic. If bad luck played some part in its poor results, then it's not surprising that they subsequently rose. The change alone does not prove the intervention worked. Put simply, when you are at the bottom, the only way is up. Equally, a school at the top of the league tables should expect to head downwards sometime soon (which may disappoint the parents who have just sent their children there).

In the statistical literature, the issue is known as 'regression to the mean' and is the pattern of unusual outcomes returning towards the typical average. It was originally known as regression to mediocrity – which is splendidly British and Victorian – and based on the observation that children of very tall parents tend themselves to be shorter (and children of very short parents, taller). It was a relationship discovered by the polymath, Sir Francis Galton (Galton, 1886). One year I was Head of Teaching in a department when we achieved a 100% approval rating from our students in a national student survey. People asked us how we did it. The honest answer was we didn't know. Just lucky, probably. And, guess what! The next year our rating went down.

The same logic has been applied to speed cameras. Evidence suggests that where there is a camera, accidents reduce (Allsop, 2013). The important policy question is whether the decline is due to the camera or whether it would have happened anyway. The statistical problem is that cameras tend only to be installed in places experiencing the highest number of accidents – so-called 'black spots'. This has a certain logic to it (why put cameras in places that haven't shown a risk?), except that the subsequent decline could be regression to the mean. Better evidence would be provided if the choice of locations was subject to a more experimental design; for example, if some of the accident black spots received cameras but some of them did not. In this way, we could compare the subsequent statistics for the two groups, which ought to show a larger decline for locations with a camera, if the cameras have an effect.

Simple in theory, but it might be difficult to explain to safety groups why only some of the dangerous traffic intersections are receiving safety measures whilst others are deliberately held back. Experimentation is difficult enough in a laboratory. Outside, in the real world, trying out actions from which only some people are seen to gain raises all sorts of controversy and ethical issues, especially when people's lives and well-being are at stake. However, their well-being is not helped by unreliable evidence either. Policy research, for example, requires knowledge of what sort of actions work best to achieve a desired outcome. To find the answers, types of experimentation (including controlled trials) are becoming more commonplace in the social sciences, just as they are used in the clinical sciences (see, for example, Dunning, 2012; Teele, 2014; Webster and Sell, 2014).

2.5 SPURIOUS ASSOCIATIONS

A lesson of the previous section is to be wary of events that coincide but are not related. Time series data provide excellent sources of spurious association. The website www.tylervigen.com gives a number of examples, including the relationship between US spending on science, space and technology, and suicides by hanging, strangulation and suffocation between 1999 and 2009 (see also Vigen, 2015). That

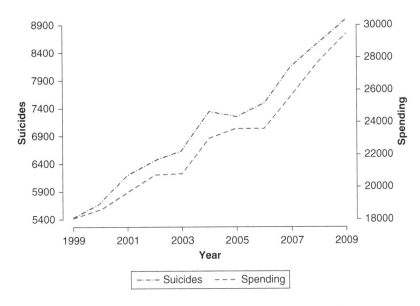

FIGURE 2.2   The number of suicides has increased at the same time as US spending on science, space and technology, but are they really directly related to each other? (source of data: www.tylervigen.com)

relationship is reproduced in Figure 2.2. The two measures move very closely together. In statistical language, we say they are correlated, and almost perfectly so.

Despite the allure of Figure 2.2, it is unlikely anyone would claim the spending causes the suicides or the suicides cause the spending. How could they affect each other? However, other associations are more seductive and less readily dismissed, especially when they fit with a preconception, prejudice or a prior point of view.

Consider Figure 2.3, which is from an academic paper questioning whether foreign aid helps or hinders economic growth (Easterly, 2003). The data are for Africa. The problem is neither with the graph nor with the article, which raises thoughtful questions. It is with others overstating what the graph conveys.

It was first shown to me as 'proof' that foreign aid does not work because it stifles economic self-sufficiency and growth. Commenting on the same chart on the BBC News website (where it appears under the headline 'Why Aid Doesn't Work'), a chief economist writes, 'if nothing else, aid to Africa seems to have lowered rather than increased economic growth'.[4] Looking at Figure 2.3, that conclusion does seem self-evident − foreign aid has increased whilst growth per capita in Africa has slumped. It could be correct. On the other hand, we might note some growth during the 1970s as aid increased, and again during the 1990s (with fluctuations).

---

[4]http://news.bbc.co.uk/1/hi/sci/tech/4209956.stm

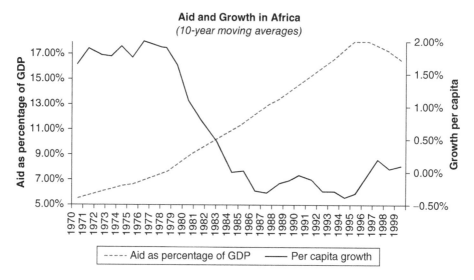

FIGURE 2.3 Exploring the relationship between foreign aid and economic growth in Africa (Easterly, 2003)

It is what happened in the intermediate period that is of real interest, when one line representing aid rises sharply and the other representing economic growth slumps. That was a period of economic restructuring and the ascendency of neo-liberal economic policies (e.g. Reaganism/Thatcherism). Perhaps it is the economic system that is the problem or the type of foreign aid being offered? Could it be a consequence of the oil price shocks during the 1970s? Maybe the death of disco precipitated the decline. The point is twofold. First, in complex systems where there are lots of potential explanations, isolating any one and establishing it to be causative is difficult (although we probably can dismiss the emergence of punk rock as a factor). Second, a coincidence of timing can be precisely that: a coincidence.

2.6 PERSUASIVE GRAPHS

Graphics are very persuasive. In a sense, that's their point – to convey a message simply and clearly. Learning quantitative methods is about effective presentation and communication, and we look at this further in Chapter 7. The trouble is that graphics can be manipulated or annotated to support a particular point of view.

Consider Figure 2.4. It appeared on the website of a national newspaper in 2013 under the heading 'The Great Green Con no. 1: The hard proof that finally shows global warming forecasts that are costing you billions were WRONG all along'. The accompanying text states: 'The graph on this page blows apart the "scientific basis" for Britain reshaping its entire economy and spending billions in

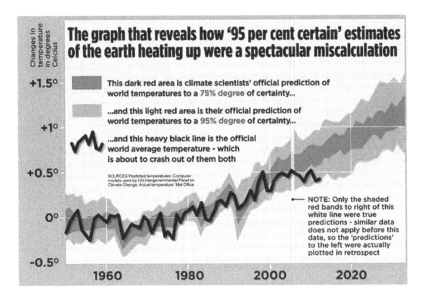

FIGURE 2.4 Be wary of annotated graphs with attention-grabbing headlines (source: http://www.dailymail.co.uk/news/article-2294560)

taxes and subsidies in order to cut emissions of greenhouse gases.'[5] The graph also was circulated amongst American climate change sceptics.[6]

Are the claims against the unreliability of global warming predictions warranted? The case is made on the basis of world average temperature falling at the lower end of official predictions and apparently heading towards falling outside of them completely. Even so, the title – 'The graph that reveals how "95 per cent certain" estimates of the earth heating up were a spectacular miscalculation' – seems wilfully misleading. Looking at the graph, there isn't a single year when the temperature is not within the 95 per cent range of estimates. Even if there were, it would not disprove the predictions. As the name suggests, a '95 per cent certain' estimate will not be correct 100 per cent of the time. In fact, by 2015, the change in temperature had risen to 1 degree Celsius, as predicted.

Reading the newspaper article emphasises the distinction between measurement and interpretation, between data and commentary. There is little sense in having data if we don't interpret them in some way, but we ought to acknowledge when those interpretations are speculative or only one of a number of possible interpretations.

---

[5]http://www.dailymail.co.uk/news/article-2294560

[6]I received it not because I am a sceptic, but because I am interested in the relationships between right-wing religious theologies and climate change scepticism in the United States. Everyone needs a hobby.

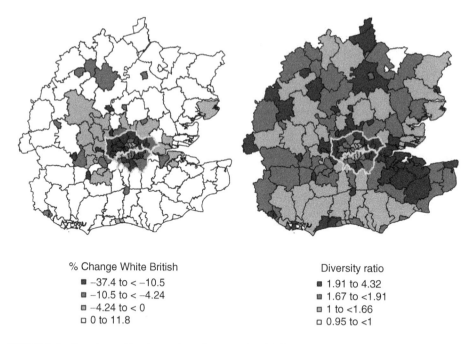

% Change White British
- ■ −37.4 to < −10.5
- ▣ −10.5 to < −4.24
- ▢ −4.24 to < 0
- ☐ 0 to 11.8

Diversity ratio
- ■ 1.91 to 4.32
- ▣ 1.67 to <1.91
- ▢ 1 to <1.66
- ☐ 0.95 to <1

FIGURE 2.5   Two views of London and South East England (left). The decline of the White British population in the capital (right). Nearly all places have become more ethnically mixed

Figure 2.5 gives an example of using the same underlying data to reach different conclusions. The map on the left shows the growth or decline in the Census count of the White British population in and around London in the period 2001–11. We may observe that in the city and in neighbouring local authorities the White British population has declined – there were fewer White British persons living in them in 2011 than in 2001. In contrast, away from the city, in more rural areas, the White British population has increased. In the media these data led to claims of 'white flight' and 'white avoidance', where the phrases imply a dislike of ethno-cultural change in the neighbourhoods left behind or a preference to live with one's own ethnic peers. However, that is only one interpretation of the data and a potentially sensationalist one.[7] The map on the right adopts a similar shading scheme but looks at the data in a slightly different way, mapping what is a simple measure of neighbourhood diversity. For this map, a value greater than 1 means the 'minority' ethnic groups are growing faster than the White British population in the location. That is the situation almost everywhere in and

---

[7]See http://quarterly.demos.co.uk/article/issue-1/understanding-white-flight/ for a more thoughtful discussion.

around London, and it means that whatever the reasons for the declining number of White British in the capital – reasons that could include a natural decline because of an ageing population or movement due to jobs, schools or family – the places where the White British are increasing have an even faster growth of other groups. In other words, neighbourhoods are becoming more ethnically mixed in and around London.

2.7 IN DEFENCE OF QUANTITATIVE METHODS

If you stopped reading at this point, you might think that the only reason to study quantitative methods is to be informed of their shortcomings. Certainly it is important to be aware of how numbers and statistics are misused, as well as what numbers reveal and conceal. Some good books on this subject include *The Tiger That Isn't* (Dilnot and Blastland, 2008), *Damned Lies and Statistics* (Best, 2012), *The Norm Chronicles* (Blastland and Spiegelhalter, 2013), *How Numbers Rule the World* (Fioramonti, 2014), *I Think You'll Find It's a Bit More Complicated Than That* (Goldacre, 2014), as well as the classic *How to Lie with Statistics* (Huff, 1991 [1954]).

However, it would be odd if the only purpose of this text were to cultivate scepticism. Odd, and counter-productive too. Used effectively, quantitative methods are important tools for obtaining evidence, testing theories, challenging the taken-for-granted and contributing to social, scientific and policy debate.

A good example of this is Thomas Piketty's somewhat surprising bestseller *Capital in the Twenty-First Century* (2014). I say 'surprising' because the hardback edition is a 685-page text, translated from French, looking at the long-term evolution of inequality, the concentration of wealth and the prospects for economic growth through the lens of economic theory. Not the most obvious candidate for holiday reading, but it reached number one in the *New York Times* hardcover best sellers in the week of 18 May 2014, remaining in the top ten until mid-August.

In his book, Piketty seeks to explain the trends shown in Figures 2.6 and 2.7, which are of decreasing social inequality for a period after World War II but rising inequality thereafter. There are several reasons why Piketty's work is a good example of using numbers and data effectively.

First, it provides evidence rather than relying on supposition. Piketty (2014: 2) notes that 'intellectual and political debate about the distribution of wealth has long been based on an abundance of prejudice and a paucity of fact'.

Second, Piketty and his colleagues had to gather the data before they could use them. This is no minor undertaking. Often the processes of gathering data and making them fit for purpose take longer than their subsequent analysis.

New technologies make the task easier, as they also facilitate easier access to data. However, off-the-shelf, ready-to-use data sets are rare, which is why having the computational skills to compile and clean data is as important as having statistical knowhow.

Third, there is a high degree of transparency to Piketty's work. All the data, together with technical notes, are available on a website and open to others to inspect, download and reanalyse.[8]

Fourth, although the inside cover has a photograph of the author set against some formidable looking equations, the book itself uses simple charts and statistics to support the economic arguments. There is nothing at all complicated about Figures 2.6 and 2.7. The first shows what percentage of the national income belonged to the most wealthy, the top 10 per cent, in the United States from 1910 to 2010. From about 1980 onwards, their share of the national income rose. The second considers how many times greater private wealth is than annual national income in selected European countries from 1870 to 2010 – about seven times in 1870 (700%), twice to three times in 1950, and rising back to about five times in 2010. Simple statistics can be extremely effective in supporting an argument, in this case one of growing inequality.[9] The expertise is in knowing how to deploy them.

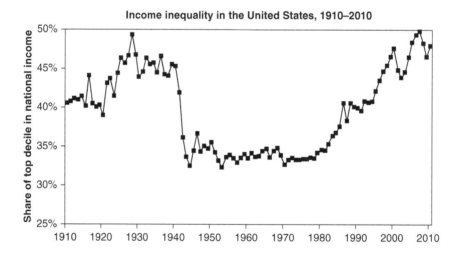

FIGURE 2.6 A graph from Thomas Piketty's *Capital in the Twenty-First Century* (source: http://piketty.pse.ens.fr/en/capital21c2)

---

[8]http://piketty.pse.ens.fr/en/capital21c2
[9]Plenty of other authors agree. See, for example, Atkinson (2015), Dorling (2014), Stiglitz (2013) or Wilkinson and Pickett (2010).

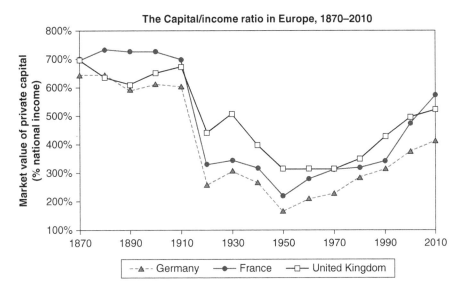

FIGURE 2.7 A second graph from Thomas Piketty's *Capital in the Twenty-First Century* (source: http://piketty.pse.ens.fr/en/capital21c2

2.8 CONCLUSION

As I wrote this, on a rainy public holiday, an item in the newspaper caught my eye. It cited a report from a children's charity, Barnardo's, claiming that one in five families cannot afford a traditional day trip to the beach. According to their research,

> the poorest families have too little money to cover basic weekly living costs – let alone a trip to the beach. The incomes of the UK's poorest families have declined in recent years. They have been hit hard by a toxic mix of rising living costs and working and non-working benefits cuts. Welfare reform has included measures that break the link between benefits and inflation.[10]

Whatever you think of this, and whether you agree with the political commentary, this is a good example of using simple statistics to produce a compelling narrative, grab the media's attention and stoke debate. That they timed the press release with the public holiday is a masterstroke.

In many respects, this is what learning quantitative methods is all about: having the ability to tell stories with data, yet also using the ability to ask intelligent questions of

---

[10]http://www.barnardos.org.uk/news/Poor_families_wave_goodbye_to_Bank_Holiday_beach_trip/latest-news.htm?ref=98360

the stories other people are telling with numbers and statistics. Let me leave the final word to Thomas Piketty whose work I cited above:

> Social scientific research is and always will be tentative and imperfect. It does not claim to transform economics, sociology [or geography] into exact sciences. But by patiently searching for facts and patterns and calmly analysing the economic, social, and political mechanisms that might explain them, it can inform democratic debate and focus attention on the right questions. It can help to redefine the terms of debate, unmask certain preconceived or fraudulent notions, and subject all positions to critical scrutiny. (Picketty 2014: 3)

The case for using data, carefully, to provide evidence for or against an idea or theory, to generate debate, to challenge the *status quo*, to argue for change, to enlighten policy or to guide decision-making, applies equally to physical and environmental geography as it does to human geography and the social sciences.

PART 2
FOUNDATIONS OF QUANTITATIVE GEOGRAPHY

3

PRINCIPLES OF STATISTICS (OR, HOW STATISTICS WORK)

3.1 INTRODUCTION

Chapter 2 talked a lot about statistics, but did so without defining what that word means. It instead relied on an everyday intuition that understands statistics as numbers, tables and graphs published in the media, in policy papers, in academic journals and in business reports. A simple example of this sort of statistics is found on the US Census Bureau's website where it lists the ten most populous countries as of 1 July 2014 (as shown in Table 3.1). To say China had a population of 1,355,692,576 is to report a statistic – a so-called fact. But it's not really a fact at all: it's a best guess, a population projection, based on the 1990 Chinese Census, the 2005 Chinese micro-census and some other data about fertility and migration. It is very unlikely that the Chinese population was exactly 1,355,692,576 at any time on 1 July 2014. The value is indicative and by no means certain.

Recognising the uncertainty, yet assuming the Census Bureau has done its best to produce a meaningful number, yields a second understanding of what statistics is about: the processes of using what are (and always will be) imperfect data to produce meaningful information, what was described as evidence in Chapter 2. A textbook introduction to statistics will focus on these processes. It will not provide page after page of numbers and tables in the style of the *Statistical Abstract of the United States*, an almanac describing social and economic conditions in the United States.[1] Instead, the textbook will introduce the methods and techniques used in the collection and analysis of quantitative data, and the thoughts and thinking behind them. It will talk about *t*-tests, *p*-values, regression and other strange vernacular.

---

[1] The *Statistical Abstract* was published annually from 1878 to 2012, when it fell victim to fiscal cuts.

TABLE 3.1   The ten most populous countries in 2014 according to the US Census Bureau (source: http://www.census.gov/popclock/)

Rank	Country	Estimated population
1	China	1,355,692,576
2	India	1,236,344,631
3	United States	318,892,103
4	Indonesia	253,609,643
5	Brazil	202,656,788
6	Pakistan	196,174,380
7	Nigeria	177,155,754
8	Bangladesh	166,280,712
9	Russia	142,470,272
10	Japan	127,103,388

There is, then, a distinction between statistics as the outcome of data collection and analysis, and statistics as the tools and know-how for generating those outcomes. It is the second of these two meanings that this book is most concerned with. If statistics are a cookie, then the interest is in the baking as well as the tasting. It's important to have some knowledge of the recipe; how else can you know if the product is good to eat?

The purpose of this chapter is to introduce some principles of statistics – a sort of top ten of 'cooking tips'. These are given in a general sense without reference to specific statistical tests and with only the slightest use of mathematical notation. I may be an outlier – statistical speak for an unusual case – but I find the concepts and ideas behind statistics interesting. I find them more interesting than the formulae used to express them. In my view, it is more important to get your head around the underlying principles than to worry too much about the equations. Let's begin with some ideas and leave the maths to the next chapter.

3.2 TEN PRINCIPLES OF STATISTICS FOR QUANTITATIVE GEOGRAPHY

Principle 1: Good data make for good analysis

The first tip is obvious but easy to forget. You can't create fine dining from poor ingredients. If what you wish to study is poorly measured then no amount of statistical wizardry can prevent the conclusions drawn out of the data from being shaky and potentially wrong. In measuring well, I don't only mean being careful when taking readings from some measuring instrument (although that is important too). I mean, more especially, that the data should be representative of what is being studied and not biased towards particular aspects of it.

Imagine wanting to know the average height of some landscape. In fact, that average is impossible to determine exactly by any form of survey. Land height is an example of something that differs continuously across a region. Even parts that are close together are unlikely to have *exactly* the same height. To determine the true average would require a measurement at absolutely every location across the region. Unfortunately there is a never-ending number of those locations because the land can be repeatedly subdivided into places that are one metre apart, one centimetre apart, one millimetre, one micrometre, and so forth. Of course, it would be a very precise measuring device that could distinguish locations one micrometre apart and look at differences in their height. Nevertheless, the problem remains: to know the true average requires a measurement at every possible location.

Assume we could measure everywhere. To do so is not a good use of anyone's time if locations that are close together tend to have similar heights. It would be sensible, faster and cheaper to sample at a fraction of the locations and estimate the average based upon those, assuming we are willing to accept that the sample average is indicative of but not exactly equal to the true average, and that true average will remain unknown. We also need to choose the locations carefully. We'll get a misleading answer if we sample only at the peaks or only in the valleys.

How about something we might sample completely? For an investigation of social attitudes we could, in principle, ask the entire population, but that is not necessary to still get a good sense of how attitudes vary by age, by gender, by income or by an ethno-cultural grouping. We don't need to reveal the complete picture to gain an understanding of what that picture is. However, we are reliant on the data being collected in a way that avoids a distorted view.

An important question is how many measurements to take. Intuitively, more is better because it will provide more information and will be less influenced by the occasional rogue result. Whilst that intuition is correct, diminishing returns set in. Stated broadly, the value added from collecting more data increases more slowly as more data are acquired.

One week before the referendum on Scottish independence, a poll for a Scottish newspaper reported that the no to independence campaign was in the lead, with 53 per cent expected to vote no and 47 per cent yes. This information was based on a survey of 1000 people. In sampling only some of the registered voters, the pollsters were accepting that their data were unlikely to yield the exact same result as if everyone had been asked. Consequently, the poll had a margin of error, and the figures were actually in the range of about 50–56 per cent no and 44–50 per cent yes. Such margins of error are like the measures of uncertainty we saw around the rankings of schools in Chapter 2 (Figure 2.1), and also like the lowest and upper estimates of maternal mortality that we shall see in Figure 3.1 later in this chapter. Would asking a few more people reduce the margin of error? Yes, but not by much. Asking 1500 or 2000 would cost more, but not enhance the estimates greatly. Asking 10,000 or 100,000 would yield improvements but the gains do not warrant the cost (especially as any published poll is immediately out of date in a fast-changing political landscape). To put it crudely, 1000 people will do.

Interestingly, it would also do if you were surveying the larger population of England (an estimated 53.9 million in mid-2013, compared to 5.3 million in Scotland) or even that of the United States (319 million). This is because the size of the required sample does not depend on the size of the population. Instead, it depends on how variable the population is and on the margin of error you are prepared to tolerate.[2] What matters most is who has been surveyed and whether they are representative of the population from which they are drawn. You would not, for example, survey only English people in Scotland and assume they represent the whole country's attitude towards independence, or survey only Democrats if you want to understand American opinions on government-funded health care.

Principle 2: Your results are sample-dependent, they depend on the data

This is just an extension of Principle 1. Classic methods of sampling work to try and avoid systematic bias towards particular people or places. Quotas may be used to ensure, for example, that men or women, or majority and minority ethnic groups, are surveyed in sufficient number to get a meaningful sense of their views, sometimes oversampling smaller population subgroups to ensure the information about them is not too scant. Oversampling means surveying a group at a rate disproportionate to their number in the overall population: they might comprise 1 per cent of the population but 5 per cent of the survey. The consequence of oversampling will be to make the sample unrepresentative of the overall population because the sample now has more of the oversampled group in it than would otherwise be expected. The solution, common to many survey data sets, is to specify weights to be used in the analysis of the data. For instance, groups that appear twice as frequently in the sample as they do in the population are given a weight of one half. That might sound a little odd – after all, if these people only count half, why bother to interview more than is required? The answer is they *are* required. If you do not increase their number then the few you do select could be very atypical of the group as a whole, the variation within the group could be

---

[2]A larger population may, of course, exhibit greater diversity (greater variability), in which case a larger sample size is required to retain the same margin of error.

underestimated or maybe the group would be overlooked entirely. Any of these circumstances could lead to misleading information and conclusions.[3]

Even if the sampling is designed to ensure an appropriate balance of males and females, of majority and minority ethnic groups, of rich and poor, of the highlands and lowlands, or whatever it is being surveyed, the exact choice of person (or location) is rarely known ahead of their selection. Usually there is a random element to sampling, so that whoever is included is not predetermined in advance but chosen by the equivalent of a toss of a coin or the roll of a die. The idea is that this helps prevent unconscious or other biases creeping into the data. Imagine I could put all the readers of this book into a bag, shake it up a little, and then reach in and randomly select one in ten. Assuming I wasn't arrested, and further assuming I do have some readers, then I can expect those removed from the bag to be fairly typical of the group as a whole.

We say that the sampling is unbiased if, hypothetically, I could put everyone back in the bag, undertake the sampling process over and over again and, in doing so, build up a picture of my audience that is centred upon who they actually are. Unfortunately, even the best efforts to ensure a sample is unbiased cannot prevent the possibility that any one sample is strange and misleading: I could reach into my bag and select all males, or all females. Those chosen could loathe mathematics. Alternatively, they may have solved the map colour problem overnight.[4] Each case is just the luck of the draw.

Problems arise if the sample does not represent the reality it is supposed to portray. Leading up to the 2015 UK general election, poll after poll suggested that the two main parties were tied for vote share at about 33–34 per cent each. The final result was that the Conservatives achieved 36.9 per cent and Labour 30.4 per cent, which gave the Conservatives a majority in Parliament.[5] It could be argued that the result was within the margin of error of the polls, but the fact that so many

---

[3]It's a bit like when popular marathons have 'good for age' entries. If registration for the marathon was only by random ballot there would be few finishers between the elite runners pushing towards 2 hours and those with a more typical average of about 4 to 4.5 hours. To fill the gap, the good for age are, in effect, oversampled by raising their probability of selection (to one, if entry is guaranteed).

[4]The map colour problem was to prove that any map with a series of areas that do not overlap and that have no gaps between them requires only four colours to have no two areas shaded with the same colour either side of their shared border. It was fiendishly difficult to prove but was done, in 1976, using computers. Arguably it changed the nature of mathematical proof because it is not reproducible 'by hand' – see Robin Wilson's *Four Colours Suffice* (2003). Cartographically speaking, it is of limited practical use.

[5]The UK system is not a proportional system but a series of 'mini-elections' for geographically based constituencies all across the country. With 36.9 per cent of the votes the Conservatives nevertheless won 50.9 per cent of the constituencies.

were centred on a tie suggests there was a systematic difference between what was being reported and people's actual voting intentions.

Principle 3: It varies

The 'it' is whatever is measured. It is, for example, the number of people in the world's ten most populous countries in Table 3.1. More generally, when we have a collection of measurements, they may be referred to as a variable, for the very good reason that they are not all the same. They may also be described as a random variable. This doesn't mean the measurement is due purely to chance or that absolutely any value is possible (negative population isn't). It just means that the value is not fixed in advance and has a random component.

Randomness can arise from the sampling, but even if we sampled the same person over and over again, we still should not expect a constant set of results. Think about if someone were to measure your height. Most likely the measurement would be broadly representative of your true height. However, it is determined by other things too: whether you happen to be standing straight at the time of the measurement, the time of day (since your height changes a little during the day), whether the person is level with the tape measure when the reading is taken, whether the tape has stretched at all, the thickness of your socks, and so forth. It is these uncertainties in the measurement that can be considered random. It is not to say they don't have causes or explanations, because in principle they do. It's just that they go into a complex mix of things that affect the measurement and mean the outcome is variable: sometimes you hunch a little; sometimes you wear thick fluffy socks; sometimes the person looks up to take the measurement. There's an element of chance to all this. Randomness means the person could measure your height a second time and not get the exact same result.

Another word for random is 'stochastic'. A stochastic process is one that has an uncertain outcome. Of course, the population values in Table 3.1 are not entirely random. There is an obvious explanation why countries like China, India and the United States have large estimated populations. It is because they are large countries and a lot of people live in them! Even so, we may still want to regard the population counts as a random variable. One reason is they make use of random surveys of the populations in their estimates. But even if the data were drawn from a complete census of the countries' populations on 1 July 2014, there would still be a stochastic element to them. For example, accidentally forgetting to deliver the census form to one of the streets; a household that happens to be out of the country on the day of the census; someone who is sick; a person who hates governmental intrusion and decides to lie; a family whose baby is born a second or two before the day ends and after they've submitted their information. These are all chance events that create uncertainty in the data.

Principle 4: The signal and the noise

Principle 3 suggests that some of the variation found in data is caused by the processes of measurement and of sampling, with the hope that these sorts of variations are random and unbiased. However, these are not the only sources of variation. If I took repeated measurements not only of your height but also of a larger sample of students in your class, the different values I would find would not be due solely to chance but would have an explanation – some people are taller than others for reasons of diet, genetics, and so forth.

The fourth principle is that not all variation is due to happenstance, and some has causes we are interested in exploring. In a theme park, daily sales of ponchos are linked to the amount of rain that fell. River flooding is caused by heavy rainfall but also the size of the watershed, its vegetation cover, geology, slope, the hydraulic radius (shape) of the river channel, how much rain there has been recently, how saturated the ground is, and vegetation growth within the river. In a system of school choice, the distance pupils travel from home to school is a function of the number of schools they can choose from, how easily they can reach them, the quality and prestige of the schools locally, the competition for places, the type or ethos of the schools, their attractiveness to the pupils' parents, and the admissions criteria the school employ. We use statistical methods to help examine the variations, to consider their causes and to learn about what is related to what else.

Sometimes statistical texts talk about signal and noise. In fact, that is the title of a book described by the *New York Times* as 'one of the more momentous books of the decade'.[6] The analogy is to a radio wave. The signal is what we want to hear – the news from the data, free of bias and interference. The noise is what degrades the signal and makes it harder to interpret – missing data, errors and so forth. Statistical methods are about tuning into the signal, exploring what it is saying and looking for plausible explanations for it.

Principle 5: The difference between a sample and a population

When thinking about statistics, always keep in mind the difference between a sample and a population. Put simply, the sample is the set of measurements we actually have. The population is the (hypothetical) set of all possible measurements we would have if we could measure everywhere or survey everyone. The population

---

[6]The *Signal and the Noise* by Nate Silver (2013) – he's the founder of fivethirtyeight.com, a website that uses statistical analysis to tell compelling stories about politics, sports, science, economics and lifestyle. If his was one of the more momentous books, I wonder what the most momentous is? Probably not the one you are reading.

need not be people, though. It is the larger group or 'thing' from and about which we are taking measurements.

Arguably the population is unknowable in a complete sense. Even a complete census has information that is missing and imperfections in the data due to stochastic errors. We have discussed already why surveying the entire population is unnecessary, even if it were possible to do so. Aside from the cost and other practicalities, in some circumstances a complete survey destroys the very thing we want to understand. That is true of food testing: if you were to remove and take a bite from every item on the production line, you would (a) be very full and (b) have nothing left to sell without teeth marks.

What we learn about the population is what we learn from the sample. That is why we want the sample to be as representative of the population as possible. Having said that, you might reasonably query the logic: we are asking for the characteristics of the sample to accurately reflect the characteristics of the population, yet we can only know the characteristics of the population in so far as we measure them in the sample. There is circularity here. How can we know the sample is representative of the population if the population is ultimately unknowable?

It's a good question, with two answers. First, we try and collect the data to minimise the possibility of bias, which takes us back to the issue of sampling well (Principles 1 and 2). Second, all things being equal, we have more confidence in larger samples of the population that we do in smaller ones. Nevertheless, the 'all things being equal' has three caveats. The first is the issue of diminishing returns. Once the data set is of a reasonable size (a few hundred or thousand, for example) then it has to be increased quite substantially to see a large return in the additional confidence we gain from it. [7] Second, you can increase the amount of data as much as you like, but if the methods of collection render it biased and inaccurate you just have more biased and inaccurate data. Third, if the sample produces highly varied measurements then that will make it harder to predict what the population is like. The logic of this is if the sample varies a lot then so, presumably, does the population from which it was drawn, making the characteristics of that population harder to pin down.

'Statistical inference' is the name given to the idea of drawing information about the population from what we know about the sample. Recall that for the Scottish independence referendum, one survey said 53 per cent of respondents would vote no. It would have been silly (or very bold) to claim exactly 53 per cent of respondents would vote against. Allowing for the pollsters' confidence in their data, the inference was that between 50 and 56 per cent of voters would vote no. That prediction was, in fact, correct. The final referendum result was 55.3 per cent voting against independence.

---

[7] Broadly, the gain in precision increases with the square root of the sample size, so it takes a large increase in the sample size to see much narrowing of the margins of error.

Principle 6: Look to the centre and spread

A small number of data are easy to write down and make sense of. As I write this, and as of about 15 minutes ago the FTSE 100 index, a measure of the share value of the UK's largest companies, was at 6400.78, down from its opening value of 6419.15, a loss partly caused by a large supermarket chain overstating its profits by £263 million ($422 million).[8]

What if you have more data to make sense of? Imagine I am planning a Christmas break to Seattle but have heard that it's not only the coffee that is wet. To get an idea about how wet, I went to the National Centers for Environmental Information (formerly the National Climatic Data Center) website and calculated the daily amounts of precipitation recorded at Seattle Tacoma International Airport for each day in December over the ten-year period from 2003 to 2012.[9] That gave me 310 measurements, more than I can easily make sense of without a summary. Therefore I calculated an average measure, the mean, which is 4.6 millimetres (equal to 0.18 inches). That's less than half a centimetre and doesn't sound too damp.

The average represents the centre of the data. If asked to predict how much rain there is on a typical December day at Seattle Tacoma Airport, the average is a good value to call. It's unlikely to be exactly right, but it's also unlikely to be wildly wrong. That said, I would be unwise to rely too much on the average alone. In fact, over the period of the data, the daily precipitation has ranged from zero (none) to 101 mm (4 inches) per day. I could plan for half a centimetre of precipitation and get exposed to twenty times that.

Consequently, I am interested not only in the central value but also in the spread of data around it. In other words, I am interested in how variable the data are. This can be measured in various ways, but an important measure is the standard deviation. If the standard deviation is low, the amount of precipitation is fairly consistent from day to day. If the standard deviation is high, the amounts are more variable. The standard deviation of the December precipitation data for Seattle Tacoma is 9 mm, around the mean of 4.6 mm. The standard deviation for the same period at Las Vegas McCarran International Airport is 3 mm, around the mean of 0.58 mm. Vegas is drier (see the mean) and consistently drier (see the standard deviation), which helps explains its water shortage problems and the shrinking water levels of Lake Mead.

The mean average and the standard deviation are two of the most important statistics. They are used to summarise data. They are used as the basis for inference. They are used to see if the variation around the average for one variable is related to the variation around the average for another. We shall meet them both again throughout the course of this book, especially Chapter 5.

---

[8] I said in Chapter 2 that it matters if you misreport a company's profits. What I didn't know when I wrote those words was that they would prove to be prophetic.

[9] http://www.ncdc.noaa.gov/

Principle 7: Presentation matters

If statistics are the outcomes of data collection and analysis, then an important part of the statistical process is the communication of results. This involves more than simply writing down numbers or producing numerical summaries; it requires consideration of how to present the data or the analysis in ways that best aid the reader to understand the take-home message. I am of the view that sloppy presentation suggests a lax approach to analysis. I am more inclined to believe the analysis is thorough and rigorous if the care for detail extends to the write-up. In Chapter 7 we shall consider some principles of presentation. Here it is sufficient to note that tables are better used when it is important to record the exact numbers, and that charts are better when the aim is to help the reader make sense of the numbers and their key take-home messages (see Few, 2012). Maps are a particular type of chart used to reveal spatial patterns and spatial relationships.

A number of commentators have offered guidance and guidelines about how best to present data honestly and effectively. These will be considered further in Chapter 7, but it all basically reduces to common sense: don't allow the presentation to distort the data, don't produce graphs that are more complex than the data require, label axes, include a legend where necessary (to explain what any shadings or symbols mean), and avoid pie charts.[10]

Other simple considerations are to include the units of analysis. For example, if you are measuring in kilometres, then say so (e.g. the driving distance from Times Square, New York, to the White House is 371 km). Don't give the data a spurious level of precision. The Japanese Census count of population as of 1 October 2010 was 128,057,352, of which females numbered 65,729,615.[11] It would be somewhat daft to suggest that females constituted 51.32826 per cent of the population. Mathematically it is correct, given the numbers (65,729,615/128,057,352 × 100), but it implies a high degree of faith in the Japanese population having been correctly and exactly counted. A value of 51.3 per cent will suffice as a reasonable guide to the gender balance.

Finally, data summaries and presentations are not just an end point of analysis, used solely to communicate results and findings. They often are part of the process of doing the analysis – of interacting with the data, querying them, making sense of them, looking for errors and omissions, testing and forming ideas. Sometimes this takes place under the banner of scientific visualisation, which is using graphics, and especially computer graphics, to understand, illustrate and gain insights into data.

---

[10]Pie charts are something of a *bête noire* amongst those interested in effective presentation of data. We shall see why in Chapter 7.

[11]http://www.stat.go.jp/english/data/kokusei/pdf/20111026.pdf

Principle 8: Probability matters

Probability is not a subject I found especially interesting at school. It is quite tedious to show that if you flip five coins there is a 10 in 32 chance three of them will land heads up. It is equally uninspiring to determine there is a 25 in 216 chance of scoring 9 if you roll three dice.

However, probability is useful because it helps to understand whether a particular event is unusual or not. There is a professor at the University of Cambridge with a website that tries to make sense of chance, risk, luck, uncertainty and probability.[12] Its animations show you need to travel about 250 miles by car to have the same risk of fatality as travelling 17 miles walking, that horse-riding and skiing have the same risk of death as each other but hang-gliding is about 16 times riskier, and that the risk of maternal death during childbirth in the United States, although rare, is higher than it is in the United Kingdom, which, in turn, is higher than in Sweden (see Figure 3.1).

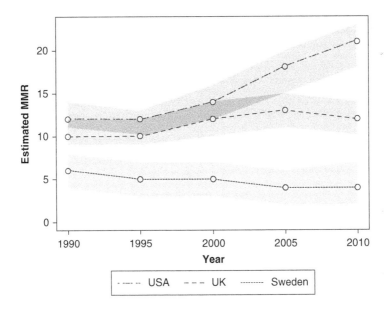

FIGURE 3.1 Estimated maternal mortality rate (maternal deaths per 100,000 live births) in Sweden, UK and USA. The shaded areas show the lower and upper range for each estimate (source of data: UN Mortality Estimation Inter-agency Group, http://www.maternalmortalitydata.org)

---

[12]The website is http://understandinguncertainty.org, and the professor is David Spiegelhalter, joint author of *The Norm Chronicles: Stories and Numbers about Danger* (Blastland and Spiegelhalter, 2013).

We need probability to help decide whether something is as surprising as it seems. A somewhat hackneyed example is to go around a class and ask everyone to share the date of their birthday, ignoring the year. Typically you only need to ask 23 people to find two with a birthday in common. Intuitively that seems low, but the intuition is wrong. Probabilities are written on a scale ranging from 0 (will definitely not happen) to 1 (definitely will). With 23 people, the probability that two or more share a birthday is $P = 0.506$, over halfway along the scale and therefore more likely than not. It's practically certain with 58 people ($P = 0.992$, over 99 per cent certain) and guaranteed if you ask 367, even allowing for leap years.

Suppose you are stuck in traffic on a four-lane freeway. You notice that the other lanes seem to move faster than yours. Isn't that just typical? Well, yes, but not exceptional: what your brain could be registering is that *any one* of the other lanes is moving faster than yours, most of the time. Tiresome if you are in a hurry, but not unfair. With three other lanes beside your own, one of them ought to be moving faster than yours about three-quarters of the time. It's only if your lane is consistently moving slowest that you can curse your bad luck for being in it.

Here's something more profound. Imagine you get diagnosed with a life-threatening disease that affects one in 10,000 of the population. The test hardly ever misses. Let's assume that it always detects people who have the disease and that it rarely gives false positives. Only 0.1 per cent of the time does it say someone has the disease when, in fact, they haven't. Is it time to make your peace with God and say goodbye to your loved ones? Let's do the maths: in a population of 1 million, 100 (one in 10,000) are expected to have the disease. That leaves 999,900 of whom almost 1000 (999.9, or 0.1 per cent) are expected to test positive without the disease. It means the group who test positive and don't have the disease is ten times bigger than the group who test positive and do. You wouldn't want to take it for granted, but the odds are in your favour.

Probability is used in statistical testing to consider whether the result of the test has arisen due to chance. For example, if I have two samples of data and their averages are different, can I infer from that finding that they come from different populations? Simply finding a difference is not sufficient. Recall Principle 3: random sources of variation will prevent them from being exactly the same. The question is whether they are sufficiently different that it is unlikely they are measuring the same thing. Assuming the data are not biased, we can use probability to assess how rarely the difference would arise if the samples were actually from the same population. The probability decreases with the amount of difference, increases with how variable (noisy) the data are, and decreases with the amount of data there are. If the probability is low, the sample averages are said to be 'significantly different', although that significance is only in a statistical sense and may not amount to anything of substance in the real world. This sort of statistical testing appears in almost any statistical textbook. It is, however, contentious. We shall discuss why in Chapter 6.

For now it is sufficient to note that it generally presupposes that what we have are randomly drawn samples when, in fact, …

Principle 9: Not all data are from random samples

The classic stages of doing a (quantitative) research project are as follows:

(a) Come up with a question, often called a hypothesis, which builds on or seeks to refute other work that has been done in the discipline.

(b) Think carefully about how to collect the data to answer the hypothesis, collect the data in a way that avoids bias, do the analysis, and from the analysis see whether there is evidence to support the initial hypothesis.

This is all very well but implies that every time a project is undertaken, new data need to be collected for the purpose. In reality, this would be time-consuming, expensive and wasteful. In practice, data are shared and used for purposes other than their original task. For many research projects the challenge is not to collect new data but to match the aspirations of the project and what it hopes to achieve with the data that are already available.

Secondary data are those that are collected not by the researcher but by some other agency. They include census information, social surveys and administrative records. Making use of secondary data requires knowledge of any potential biases or shortcomings. Although data sets produced by governments are often reliable, they are rarely perfect. Censuses, for example, will tend to undercount the hardest-to-reach populations – perhaps the homeless, illegal immigrants, or households in properties of multiple occupancy. Classic statistical tests assume the 'noise' in the data is random variation around the 'signal'. It follows that any missing data ought to be missing at random, without systematically excluding any one group, and that any errors or inaccuracies should be neutral overall.

The credibility of these assumptions is stretched to breaking point by the data used in some social and scientific research. This is especially so as classic sources of data such as censuses and traditional surveys are challenged by other ways of generating information such as the linking of administrative records or by commercial information collected by companies and used in marketing. We live in a data-rich world where data are generated in all sorts of ways, including through social media. Researchers at the Centre for Advanced Spatial Analysis at University College London investigated patterns of Twitter profanity by monitoring tweets and mapping the results.[13] They discovered that people in the North Yorkshire seaside resort of Redcar were most likely to swear. Does that mean people who live in

---

[13]http://www.bbc.co.uk/news/uk-29124416

this area are the most foul-mouthed in Britain? Possibly. On the other hand, it is probably safe to assume that users of Twitter on a smartphone app are not representative of all people, many of whom don't own a smartphone and have little interest in communicating to others in 140 characters or less. In addition, all we know is that the phone was in Redcar at the time of the tweet. Quite who the person was who sent it or where they came from is unknown (#biasedsample?).

Nevertheless, even unconventional data can be revealing. A well-known search engine provider reckoned it could detect a disease outbreak faster than conventional authorities by seeing an increase in people searching for information about it. People who have access to a computer are not fully representative of the wider populace (especially in less economically prosperous regions), but the information remains useful, albeit that the reliability of the approach has since been questioned.[14] Not everyone has a mobile phone either, but the WorldPop data sets for West African countries look at population distributions and mobility patterns using mobile phone data to support efforts in controlling the Ebola virus outbreak.[15,16]

Principle 10: Geography is awkward (but interesting)

Figure 3.2 shows the average house price for properties sold in London between 2011 and 2013. There are a number of comments to make. First, house prices in the capital are really high. Second, by mapping the data we reveal their geography — that's why maps are useful. The most expensive parts of London are in two sectors, extending from the centre to the northwest and to the southwest. The cheapest are in clusters to the west, south, north and, the largest cluster, to the east. Third, it seems unlikely that the patterns in the map are generated by chance. More probably, they reflect the socio-economic geography of the city.

In fact, what the map exhibits are patterns of spatial dependency and spatial heterogeneity. 'Spatial' is essentially a synonym for geographical, 'dependency' means that the house prices in one place are influenced by the prices of other places (most noticeably the places nearby, which is why clusters form on the map) and 'heterogeneity' just means diversity: what we see in one part of the map isn't necessarily what we find elsewhere. From a statistical point of view, these patterns

---

[14]See http://www.nature.com/news/when-google-got-flu-wrong-1.12413 and http://time.com/23782/google-flu-trends-big-data-problems/

[15]http://www.worldpop.org.uk/ebola

[16]On the day I wrote this, I thought it might be interesting to see what was trending on Twitter in my home city of Bristol. It was the September Apple Live event from two days before. I then thought it might be interesting to compare the trends with those for New York. Eight out of the top ten were sadly poignant. It was only when I saw them that I remembered the date: 911anniversary; September11; NeverForget; Remember911; RIP911; World Trade Center; R.I.P; and Twin Towers.

are awkward. Many statistical tests, including those regularly used by geographers, assume the data values are independent of each other. Independence ensures that the outcome of a coin toss in northwest London is not influenced by the outcome of a coin toss in southwest London, but can the same be said of house prices? Probably not: as house prices rise in one area it is reasonable to assume there is an inflationary impact upon surrounding areas too. That connection, or interdependency, will violate any statistical assumption of independence.

Geography is awkward. It also is interesting. Think how dull the map would be if everywhere were the same. I am not suggesting it is a good thing that one part of London had an average property price of £3.8 million, which is more than 180 times greater than half the population of the UK earn in a year. Rather, it would seem to hint at gross inequalities within society and the different opportunities that people can afford. That's the real point: geography reveals something useful which can form the focus of further study. What are the reasons why the property prices are higher in some places than in others? Can the differences be explained

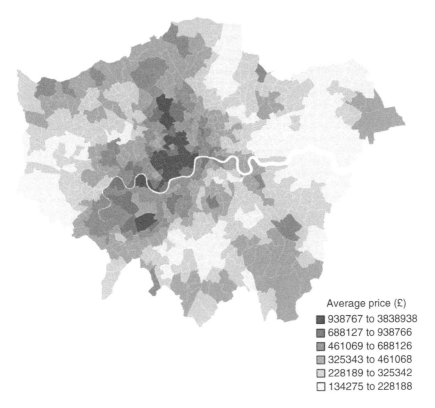

Average price (£)
- 938767 to 3838938
- 688127 to 938766
- 461069 to 688126
- 325343 to 461068
- 228189 to 325342
- 134275 to 228188

FIGURE 3.2 Average selling price in London's neighbourhoods, 2011–13 (source: Calculated from Land Registry UK data)

by the size of the properties sold? Is it to do with the age of the properties or access to amenities? Is it because particular parts of the city historically are valued as good neighbourhoods or because of the quality of local schooling? Statistically speaking, geography can be a nuisance but, geographically speaking, it reveals the social or natural processes that are important to explore and explain. What is required is a set of more geographically minded methods of analysis. We consider some of these in Chapters 10 and 11.

3.3 CONCLUSION

This chapter offers ten recipe-card principles about how statistics work. At the risk of spoiling their flavour, they may be boiled down even further. First, all analysis is dependent on the quality of the data (as well as the quality of the analyst). That is true of both quantitative and qualitative research. Second, what we learn of what we measure is through the measurements we have, so our understandings are always sample-dependent, uncertain and incomplete – even if we think we have sampled everyone or everything, we cannot assume the data will be error free. Our hope is that those errors are random and not biasing our conclusions. Third, the (mean) average and measures of variation are key statistics that are used to sum-marise data, to undertake statistical inference and to see if one variable is related to another. They will appear in later chapters that look more closely at specific statistical methods. First, however, we consider some of the principles and language of mathematics that are relevant to quantitative geography.

4

SOME MATHS AND NOTATION

4.1 INTRODUCTION

Occasionally a question goes around social media such as $1 + 3 + 2\times0 + 1 + 2\times2 = ?$ Many people get the answer wrong. Some mistakenly work from left to right, adding or multiplying as they go. Some pride themselves on spotting a 'shortcut' – noticing the zero and wrongly ignoring everything to the left of it on the basis that any number multiplied by zero is zero. Possible answers include 5 and 6. Both are incorrect. The correct answer is 9. If you disagree, see Section 4.3 below.

Maths has rules so that calculations that ought to generate a consistent result do so. It is merely embarrassing to be wrong on a website. It would be catastrophic if engineers made a mistake when calculating the maximum load for a bridge. There was once a Geographical Information System (GIS) with an inbuilt calculator that did not respect the order in which mathematical calculations should be conducted. There are, presumably, any number of applications that used that GIS and are wrong.

This chapter outlines some of the core mathematical ideas and notation to be used in the rest of the book. It introduces some of the 'everyday' maths you will encounter in geography, which are useful for reading academic journals, conducting research and demonstrating proficiency with numbers that is helpful for the workplace beyond. There is nothing specifically geographical about this chapter. Whilst it does include some geographical examples, the main purpose is to lay some foundations upon which the rest of the book builds. It is intended to be as user-friendly as a chapter about maths will ever be. As such, it may make the more mathematically minded cringe because it is telling you things you already know, because some of the mathematical explanations are incomplete and simplified, or because the attention to some topics is cursory. However, it may also make some readers nervous, especially towards the end when matrix maths is considered. My advice is to stick with it and not to be too worried if it doesn't all fit into place on a first reading. Much if it will appear in more applied settings in later chapters. This one gives you something to refer back to.

Undoubtedly there are fuller texts that go further and deeper into the sorts of maths you might need, whether you orientate to the human or physical side of geography (see Tan, 2012; Siegel and Moore, 2013; or Parkhurst, 2006). One chapter cannot cover the same content as a book. However, it can refresh and, in some cases, build on the sorts of maths you encountered at school and will need again in your degree.

4.2 INEQUALITY AND SIGNS OF THE TIMES (THE BENEFITS OF GENERAL NOTATION)

Global inequalities of wealth have become a source of social and political concern. On 19 January 2015, news channels reported a study by the charity Oxfam predicting that by 2016 the wealthiest 1 per cent of the world's population collectively will own more than half the world's wealth (they subsequently said the prediction had been fulfilled: Oxfam, 2016). The geographer, Danny Dorling, has written extensively on inequality and its potentially corrosive effects on society (see Dorling 2010, 2013, 2014).[1] Economic inequality also is the concern of Thomas Piketty's (2014) book, *Capital in the Twenty-First Century*, as discussed in Chapter 2. That author summarises the overall logic of his conclusions with an equation, $r > g$. For a book about the long-term evolution of inequality, it is appropriate that a mathematical inequality (where one value is greater than the other) is used to describe it.

The equation states that r is greater than g. Making sense of this equation is possible only if we know what r and g refer to, since they are merely placeholders (arbitrary letters) to represent something else. In fact, 'r stands for the average annual rate of return of capital, including profits, dividends, interest, rents, and other income from capital, expressed as a percentage of its total value, and g stands for the rate of growth of the economy, that is, that annual increase in income or output' (Piketty, 2014: 25). The argument is that gains in privately held wealth outstrip national gains in wealth, benefiting only those who have money through, for example, inherited wealth, and that increasing inequality will follow unless there is greater redistribution of that wealth.

Whether you agree with the argument or not (personally, I am sympathetic), the economics is not of immediate relevance. The focus is on the way that an equation can express a concept or an idea, provided we understand the language of the equation. Even something as simple as an average is difficult to express eloquently in words. The description 'add together all the numbers and then

---

[1]Dorling's *The 32 Stops* (2013) is a tour of the extent and impact of inequality along London's Central Underground line.

divide their total by the number of numbers you had' is correct but clunky. Mathematical notation tries to avoid ambiguity. Algebra is the part of maths where general formulae and equations are written with the substitution of numbers with letters and other symbols. This can be very off-putting, especially as the complexity of the equations increases. Yet, all the equations do is define as precisely as possible the nature of a mathematical expression and state, where relevant, the limits on when it can be applied.

Let's begin with some simple maths:

$$2 \times 5 = 5 \times 2 \tag{4.1}$$

This equation says that the multiplication of 2 by 5 is equal to the multiplication of 5 by 2; so the order of multiplication does not matter. Although we might assume that the same is true of other numbers, equation (4.1) is tied to a specific example. To state the claim more generally, we need to move from the specific to the algebraic, substituting letters for numbers, as in

$$a \times b = b \times a$$

or, more simply,

$$ab = ba \tag{4.2}$$

The letters a and b are arbitrary, and simply mean 'a number'. We arrive back at equation (4.1) if we set $a = 2$ and $b = 5$. The point is, we could substitute a and b with any numbers and the equation would remain true. Equally, because the letters are arbitrary, they could be changed too. For example, the same property of multiplication could be illustrated using $xy = yx$.

What the equation defines is the commutative property of multiplication. It is true also of addition,

$$a + b = b + a \tag{4.3}$$

but not of either division or subtraction. For example,

$$a/b \neq b/a \tag{4.4}$$

which reads as 'a divided by b (or a over b) is not equal to b divided by a (b over a)'. For instance, $6/3 = 2$, whereas $3/6 = 0.5$.

We could also write equation (4.4) using the division sign: $a \div b \neq b \div a$. We might also note that a divided by b is the inverse of b divided by a. As forewarned, mathematical nuance is forsaken here. Tacitly we are assuming that a is a different number than b and that we are not dividing by zero.

For subtraction, commuting the order generates 'the same' number but changes whether it is positive or negative. For instance, $6-3=3$ whereas $3-6=-3$. We can say that both these numbers have the same magnitude, which is their value ignoring any positive or negative sign. The symbol $|\ \ |$ is used to indicate magnitude, so $|6-3|=3$ as does $|3-6|$. Using algebra,

$$|a-b|=|b-a| \tag{4.5}$$

Imagine we want to find the average deviation from the line of the points shown to the left of Figure 4.1. Because the line is in the middle of the points, the distances above it cancel out the ones below, suggesting the average distance from the line to a point is zero. This clearly isn't true: all of the points lie off the line. One way to solve the problem is to take the magnitude of how far each point is from the line and calculate the average of that. The average deviation is now found to be at 1.78 units away from the line, as shown to the right of Figure 4.1.

Returning to multiplication, it has an associative as well as commutative property. Imagine there are three numbers, such as $4\times2\times10$. We could, amongst other possibilities, work from left to right, multiplying the 4 by the 2 and then by the 10; alternatively, we could multiply the 2 by the 10 before multiplying their product by 4. The product is the answer you get when multiplying numbers together. To distinguish these orders of multiplication, brackets can be used:

$$(4\times2)\times10 \qquad \text{[multiply the 4 by 2, then multiply by 10]}$$

and

$$4\times(2\times10) \qquad \text{[multiply the 2 by 10, then multiply by 4]}$$

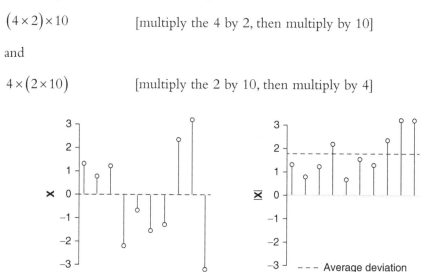

FIGURE 4.1 A set of values, collectively represented by the notation X (left). The magnitudes of the same values, $|X|$ (right)

In either case the result is the same because

$$\left(ab\right)c = a\left(bc\right)$$
(4.6)

The associative property is true also of addition,

$$\left(a+b\right)+c = a+\left(b+c\right)$$
(4.7)

but not of division nor subtraction. For example, $\left(a-b\right)-c \neq a-\left(b-c\right)$.

4.3 THE ORDER OF CALCULATIONS

It is tempting to treat an equation like a book, reading from left to right. Sometimes this will work, but with a mixture of mathematical operations (e.g. addition and multiplication) mistakes can arise like the ones seen on social media. For example, $3+2\times6$ equals 15 not 30, and $110-10\div5$ equals 108 not 20. The reason is that the order of calculation gives priority to multiplication over addition, and to division before subtraction, meaning they are evaluated as $3+12$ and $110-2$ (and not as 5 × 6 nor 100 ÷ 5).

This order can be overruled by the use of brackets, more formally known as parentheses. The calculation $(3+2)\times6$ does equal 30, because the part in parentheses is evaluated first. Similarly, $(110-10)\div5$ equals 20. Use of parentheses is often sensible even when not necessary, where it provides clarity. The answer to the introductory question is obvious if the stages of calculation are clear:

$$1+3+2\times0+1+2\times2 = 1+3+\left(2\times0\right)+1+\left(2\times2\right)$$

$$= 1+3+\left(0\right)+1+\left(4\right) = 9$$

Parentheses are often used in programming to help avoid unintended errors. When the parentheses are nested, you work from the inside out:

$$5+\left(2\times\left(3+1\right)\right) = 5+\left(2\times4\right) = 13$$

In maths, the order of operations begins with parentheses then moves to exponents, multiplication, division, addition and subtraction. As a mnemonic it gives PEMDAS: <u>P</u>odgy <u>E</u>lephants <u>M</u>ay <u>D</u>o <u>A</u>lmighty <u>Sh</u>★★s. Exponents are discussed in Section 4.5.

4.4 RUNNING AWAY WITH MURDER (PERCENTAGES, RATES AND PROPORTIONS)

Parkrun organise free, weekly and timed 5 km runs every Saturday at locations all around the world.[2] My nearest course is Pomphrey Hill. As its name suggests, it's not the flattest course, which is a bit wearying by the third lap. Last week the runners also faced some strong winds. Did those slow people down? Looking at the course results, 35 people completed the course in less than 25 minutes, which is 10 more people than the 25 who managed it the week before. On face value, more people ran faster last week, suggesting the wind did not slow them down. However, to look just at these raw numbers is not comparing like with like. There were 143 runners who took part on the windy day, and only 101 the week before. Those who completed the course in less than 25 minutes were therefore 35 of 143 on the windy week, and 25 of 101 the week before.

In mathematics, the word 'of' indicates a fraction, which is some part of a greater whole. Currently we are trying to compare 35/143 with 25/101, which is not easy because the denominator – the number at the bottom – is different for the two events. A percentage re-expresses the fraction in a way that allows it to be compared with other fractions expressed in the same way. It doesn't change the numeric value of the fraction, in much the same way as 3/4, 6/8, 9/12 have different denominators but are still all equal to 0.75. All that happens with a percentage is that the data are fixed to the same scale, as a fraction of 100.

If x represents the number of people running the course in less than 25 minutes, and N is the total number taking part, then the percentage running it in less than 25 minutes is

$$\% = \frac{x}{N} \times 100 \tag{4.8}$$

where % denotes a percentage that was, for last week, $\frac{35}{143} \times 100 = 24.5\%$. The week before, it was $\frac{25}{101} \times 100 = 24.8\%$. It seems that the wind made little difference.

Percentages can be used for statistical trickery. If N is low, then converting the values into percentages acts to inflate their importance in a somewhat misleading way. It sounds impressive to learn of the occasion when 100 per cent of runners completed the Pomphrey Hill course in less than 25 minutes. It appears more

---

[2]If you've never tried one, find your nearest at http://www.parkrun.org.uk/events/ and give it a go!

dishonest when I inform you it was me, running alone one Tuesday afternoon. Nevertheless, expressing values on a common scale is important when comparing geographical regions that vary in area and in population size. According to the FBI's Uniform Crime Reports there were 316 murders in Detroit in 2013 and 335 in New York.[3] Does that make New York the more dangerous city? Not necessarily, because it has the larger population. Expressed as a percentage of that population, the murder rate is 0.0452% in Detroit, more than ten times greater than the rate of 0.00399% in New York.

One homicide is one too many but thankfully murder is rare: both cities generate a percentage so small that the rates are better expressed in another way. One alternative is to give the number of murders as a fraction of every 100,000 of the population. Since we are now calculating out of 100,000 (instead of 100), equation (4.8) becomes

$$\frac{x}{N} \times 100,000 \tag{4.9}$$

giving a rate of 45.2 murders per 100,000 population in Detroit, and 3.99 in New York. The relative risk remains the same − more than ten times greater in Detroit than in New York − but expressing it out of 100,000 reflects the relative scarcity of a homicide.

A proportion is where the number of cases is left as a fraction of the total, without any further multiplication; that is,

$$p = \frac{x}{N} \tag{4.10}$$

Whereas percentages typically range from 0% (none) to 100% (everyone or everything), the corresponding proportions are from $p = 0$ to $p = 1$. Probabilities are measured on this scale, where 0 indicates an event definitely will not happen (e.g. rolling 7 with a conventional six-sided die) and 1 indicates it certainly will (e.g. rolling any whole number between 1 and 6, inclusive). Use of probability is a way of handling uncertainty. We do not know in advance which way a coin will land but, assuming the coin is fair, the probability it will land head up, is $P(H) = 0.5$. At halfway between 0 and 1, it is just as likely as not.

4.5 EXPONENTS AND ROOTS

Exponents take the form a^n and act to multiply a number by itself n times, where n is called the exponent. For example, 5^2 or 'five squared' has an exponent

---

[3] http://www.fbi.gov

of 2 and is equal to $5 \times 5 = 25$; 4^3 or 'four cubed' has an exponent of 3 and is equal to $4 \times 4 \times 4 = 64$. We say that the first number has been raised to the power of the second: here we have 5 raised to the power of 2, and 4 raised to the power of 3.

By convention, any number raised to the power of 0 is equal to 1:

$$a^0 = 1 \tag{4.11}$$

and a power less than 0 inverts the number,

$$a^{-n} = \frac{1}{a^n} \tag{4.12}$$

so $2^0 = 1$ whereas $2^{-1} = 1/2$, and $2^{-3} = 1/2^3 = 1/8 = 0.125$.

Where the exponent takes us from one number to another, the root ($\sqrt[n]{}$) takes us back again. The square root of 25 is 5 ($\sqrt[2]{25} = 5$). The cubed root of 64 is 4 ($\sqrt[3]{64} = 4$). The root may also be expressed as $a^{1/n}$. That is, $\sqrt[2]{25} = 25^{1/2} = 25^{0.5} = 5$, and $\sqrt[3]{64} = 64^{1/3} = 4$. More generally:

$$\text{if } a^n = b \text{ then } \sqrt[n]{b} = a \tag{4.13}$$

Where n is not specified, you can assume it is equal to 2: $\sqrt{16} = \sqrt[2]{16} = 4$. It can be helpful to remember that

$$\sqrt{\frac{a}{b}} = \frac{\sqrt{a}}{\sqrt{b}}$$

You cannot always determine the root of a number. For example, the square root of -2 is not real, because no real number can be multiplied by itself to give a negative answer.[4] In other circumstances, there may be more than one solution. For instance, $\sqrt{16}$ is actually ± 4, where the symbol $\pm$ means there are two answers, $+4$ and -4 ($-4^2 = 4^2 = 16$). Often the negative number is ignored, but whether it should be depends upon what you are trying to calculate and whether the answer could be negative.

Powers of 10 can be used to represent large or small numbers. This is sometimes referred to as scientific notation. Computers often employ this but write the notation in a slightly different way, as shown in Table 4.1.

---

[4] But a complex number can: $\sqrt{-1}$ is called the imaginary number, i, and $\sqrt{-2} = \sqrt{2}i$.

TABLE 4.1 Example of writing numbers using scientific notation

Number	Number written in scientific notation	Which is equivalent to	How a computer might show it
330,000	3.3×10^5	$3.3 \times 100,000$	3.3E5
33,000	3.3×10^4	$3.3 \times 10,000$	3.3E4
3300	3.3×10^3	3.3×1000	3.3E3
330	3.3×10^2	3.3×100	3.3E2
33	3.3×10^1	3.3×10	3.3E1
3.3	3.3×10^0	3.3×1	3.3
0.33	3.3×10^{-1}	3.3×0.1	3.3E-1
0.033	3.3×10^{-2}	3.3×0.01	3.3E-2
0.0033	3.3×10^{-3}	3.3×0.001	3.3E-3
0.00033	3.3×10^{-4}	3.3×0.0001	3.3E-4
0.000033	3.3×10^{-5}	3.3×0.00001	3.3E-5

Statistical operations often take advantage of the fact that squaring a number removes any negative sign and can solve the problem of positive values cancelling out negative ones. However, it does so at the expense of exaggerating the variation because the act of squaring gives increased influence to values furthest from the centre (see Figure 4.2). This is a problem that affects the standard deviation, a statistical measure of variation considered further in Chapter 5.

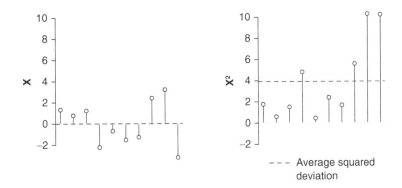

FIGURE 4.2 The same set of values shown in Figure 4.1 (left). The values squared (right)

4.6 LOGARITHMS AND EXPONENTIAL GROWTH

Imagine we raise 10 to the power of 1, 2, 3 and 4. The answers are $10^1 = 10, 10^2 = 100, 10^3 = 1000$ and $10^4 = 10,000$. The logarithm of those numbers is the power by which the base of 10 was raised to give the answer. In other words, it is the exponent. So, $\log_{10} 10 = 1$, $\log_{10} 100 = 2$, $\log_{10} 1000 = 3$ and $\log_{10} 10,000 = 4$. The relationship is more easily expressed algebraically:

$$\text{if } 10^n = b \text{ then } \log_{10} b = n \tag{4.14}$$

$\log_{10}$ is known as the common log, but any base above zero could be used. For example, $\log_5 25 = 2$ $\left(\text{and } 5^2 = 25\right)$ and $\log_4 64 = 3$ $\left(\text{and } 4^3 = 64\right)$:

$$\text{if } a^n = b \text{ then } \log_a b = n \tag{4.15}$$

Along with the common log, another regularly used logarithm is the natural log, sometimes written as $\log_e$ and sometimes as ln. The natural log has a base with value approximately equal to 2.718282. That value is known as the exponential value, e,

$$e \cong 2.718282 \tag{4.16}$$

where $\cong$ means 'approximately equal to'. The exponential value is given its own symbol because it is an irrational number, meaning it cannot be expressed as a ratio dividing one whole number by another. Another well-known irrational number is pi, which is given the Greek symbol, π, and is approximately equal to 3.141593. Many people encounter it when calculating the area or circumference of a circle (πr^2 and $2\pi r$, respectively).

Logarithms are useful for modelling exponential growth. Consider the following example:

$e^1 \cong 2.718282,$ $e^1/e^0 \cong 2.718282$ (recall, $e^0 = 1$)

$e^2 \cong 7.389056,$ $e^2/e^1 \cong 2.718282$

$e^3 \cong 20.08554,$ $e^3/e^2 \cong 2.718282$

$e^4 \cong 54.59815,$ $e^4/e^3 \cong 2.718282$

Notice that the value is getting bigger and bigger on the left-hand side of the equations. It is growing exponentially. On the right-hand side, the rate of change is constant. The same thing occurs with unpaid debt. Imagine a debt of $100 with monthly interest of 10 per cent. After 1 month $110 is owed, after 2 months $121, after 3 months $133. The debt by the end of the year is $314. In each case, the *rate* of change per month is the same: it is always 10 per cent greater than the month before. However, the amount owing grows quickly.

Figure 4.3 illustrates this. The top left graph shows the monthly increase in the debt and the top right graph shows the total debt owed. Both are curving upwards – a trend that characterises exponential growth. The bottom left graph plots the logarithm of the data and confirms what we know already: the rate of change is constant, as shown by the straight line. In this case, the natural logarithm was used. However, it doesn't matter. A straight line would be produced with any other logarithm too.

The remaining graph conveys the same information but presents it in a different way. In the bottom left graph, the log of the debt is plotted, on a standard vertical axis. In the bottom right, the actual debt is plotted but the vertical axis has a log scale – notice how the labels get closer together as you go up it. Both methods amount to the same thing. Both act to straighten the curve.

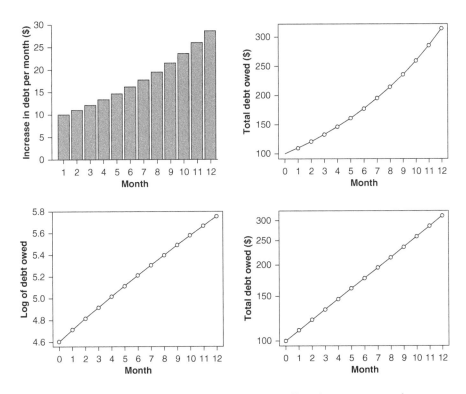

FIGURE 4.3   An example of an unpaid debt growing exponentially and at a constant rate from one month to the next

It is not just debt that grows exponentially. Cases of infectious disease can increase rapidly if people or animals are in close contact and there is no medical intervention. Animals such as mice can breed rapidly, creating large colonies and infestations. There's a slightly disturbing example on Animal Planet.[5]

In calculations involving logarithms it can be useful to know that

$$\log(xy) = \log(x) + \log(y)$$

$$\log\left(\frac{x}{y}\right) = \log(x) - \log(y)$$

$$\log(x^n) = n\log(x)$$

(4.17)

These rules work for any base, assuming that it, x and y are each greater than zero.[6]

4.7 CALCULATING A GRADIENT AND THE EQUATION OF A LINE

Figure 4.4 is a scatterplot where each dot represents a Census ward, which is a neighbourhood in London. Its position on the chart is determined by the percentage of the population that is of an ethnicity other than White British, and also by the percentage of households that are exhibiting at least one dimension of deprivation.[7] A line of best fit has been fitted to the points. It is, in fact, a regression line, a method of analysis discussed in Chapters 9 and 10. It is not a line of perfect fit – there is variation around it – but it goes through the 'middle' of the points. The line summarises a general relationship: the greater the percentage of the population that is not White British, the greater the amount of deprivation. This suggests there is an economic disadvantage associated with being from what is, in the UK, an ethnic minority background.

The gradient of the line is the vertical displacement relative to the horizontal displacement – how much it goes up or down as a fraction of how far it goes along. It is conventional to describe the horizontal axis of a graph as the X-axis, and the vertical axis as the Y-axis. If we use the Greek letter, Δ (delta) to indicate 'change in', then

---

[5]http://youtu.be/IOwinLWrEIw
[6]log(0) is minus infinity, whereas the log of a negative number generates an error.
[7]Defined as unemployment, low level of education, poor health and/or living in a shared or overcrowded dwelling, or in a property lacking central heating.

$$\text{gradient} = \frac{\Delta y}{\Delta x} \qquad\qquad (4.18)$$

which measures the change in Y relative to the change in X.

Because the line has a constant gradient in Figure 4.4, the distance over which we measure the change does not matter. For a distance of 60 units along the X-axis it rises by 22 units along the Y-axis. The gradient is $22/60 \cong 0.367$. What that tells us is that for every 1 per cent increase in the population that is not White British, we expect the percentage of households exhibiting deprivation to rise by 0.367 per cent. That is a positive relationship, but only in the mathematical sense that as one value increases so too does the other: the line slopes upwards with a positive gradient.

The value of Y for when $x = 0$ is known as the Y-intercept. If we know this value and also the gradient, then we have all the information we need to define the equation of the line:

$$y = Y\text{-intercept} + \text{gradient} \times x \qquad\qquad (4.19)$$

The line in Figure 4.4 has a Y-intercept of 40.9 and the gradient is 0.367. Therefore the equation of the line is $y = 40.9 + 0.367x$. This is more than a mathematical nicety, because it can be used to make predictions. For example, for a ward that is 50 per cent not White British in London, we substitute the number 50 for the x

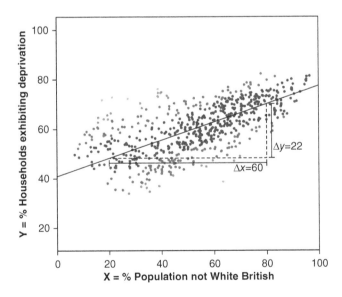

FIGURE 4.4 A line of best fit showing that there is, on average, a greater percentage of households exhibiting deprivation in London's neighbourhoods where a greater percentage of the population are from a non-White British background

and obtain an expected value of deprivation of $y = 40.9 + (0.367 \times 50) = 59.3$ per cent. It's a reasonable prediction. The actual values at this threshold range from 45.1 to 72.5 per cent of households exhibiting deprivation.

Things are more complicated if the line is not straight and therefore its gradient not constant. For example, the gradient of the curve in Figure 4.5 depends upon the value of X around which it is measured. The notation $\delta y/\delta x$ represents the instantaneous rate of change at any point on the curve. Replace δ with Δ and you will see it is the equivalent to equation (4.18). You may also recognise the notation as representing a mathematical derivative, linking the change in one variable (Y) to the change in another (X).

The equation of the curve in Figure 4.5 is

$$y = -x^2 + 200x - 200 \tag{4.20}$$

It is an example of a quadratic equation.[8] Its derivative is

$$\frac{\delta y}{\delta x} = -2x + 200 \tag{4.21}$$

The detail of the equation is not important. What matters is that the gradient, $\delta y/\delta x$, will be different for different points on the curve; that is, for different values of x.

Looking again at Figure 4.5, at the highest point of the curve the gradient is seen to be flat. That information can be used to determine the value of X at the maximum value of Y. At the maximum, $\delta y/\delta x = 0$, therefore at this point,

$$0 = -2x + 200$$

$$2x = 200$$

$$x = 100 \tag{4.22}$$

which Figure 4.5 shows to be correct.

Not all quadratic equations peak in this way. For example, the equation $y = x^2 - 200x + 10000$ produces a U-shaped curve. For it, the gradient is zero at the bottom of the curve, where the minimum value of Y is found.[9]

---

[8] Quad as in square because the x gets squared.
[9] Whether it is a maximum or a minimum can be determined by taking the derivative of the derivative and seeing whether it is negative (for a maximum) or positive (for a minimum). The derivative of equation 4.21 is –2. For $y = x^2 - 200x + 10000$, the derivative is $2x - 200$ and the derivative of that is +2.

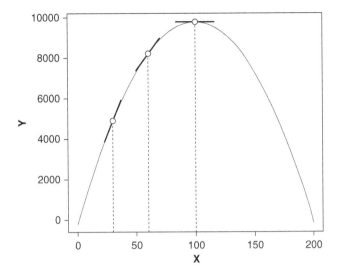

FIGURE 4.5   The quadratic relationship between Y and X produces a curve, the gradient of which changes at different points along it

4.8 CALCULATING SHORTEST DISTANCE

Figure 4.6 plots my run from home to work on a regular (Cartesian) grid, having used a smartphone to track my journey. The grid is a section of the Ordnance Survey National Grid, a system that provides map coordinates for locations in the UK. Ordnance Survey is able to use a grid system because the area of the UK is sufficiently small that the curvature of the Earth's surface will not create any major projection errors. That is handy because it makes calculating distances between locations straightforward.

How far is the University from my house? The shortest, straight-line distance is shown on the map. I run from east to west, so my starting point is at coordinate (366417,176031) and the finish is at (358036,173187). The unit of measurement is metres. The distance between the two can be calculated as

$$d_{min} = \sqrt{\left(x_{start} - x_{finish}\right)^2 + \left(y_{start} - y_{finish}\right)^2}$$ (4.23)

which is

$$d_{min} = \sqrt{\left(366417 - 358036\right)^2 + \left(176031 - 173187\right)^2}$$

and equal to approximately 8850 metres (8.85 km).

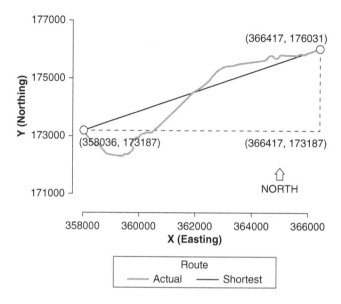

FIGURE 4.6   The route of my journey from home to work: the actual route and the shortest (straight-line) distance

The calculation is made from Pythagoras's theorem.[10] The calculation works because of the grid coordinates. It would not work if the coordinates were expressed using longitude and latitude, for which the starting point is at (51.4831°N, 2.4850°W) and the finish is at (51.4570°N, 2.6054°W). Either these need to be converted to grid coordinates or the great-circle distance calculated instead.

The straight-line distance is known as the Euclidean distance. Imagine, however, that I ran west from my house until I was due north of my office and then ran south to reach it. The total distance travelled would be the distance along the X-axis plus the distance along the Y-axis: $d = |x_{start} - x_{finish}| + |y_{start} - y_{finish}|$. This is sometimes referred to as the Manhattan distance, because it is the way you have to traverse cities that are built on a grid system with the routes running north–south and east–west.

The great-circle distance is the shortest distance between two points along the surface of a sphere (the Earth). The great-circle distance between London Heathrow Airport and John F. Kennedy Airport in New York is shown in Figure 4.7. To calculate the distance is complicated, involving trigonometry. One formula (the haversine formula) is

---

[10]For a right-angled triangle, the square of the length of the hypotenuse is equal to the sum of the squares of the other two sides.

$$d = 2R \sin^{-1}\left(\sqrt{\sin^2\left(\frac{lat_A - lat_B}{2}\right) + \cos\left(lat_A\right)\cos\left(lat_B\right)\sin^2\left(\frac{long_A - long_B}{2}\right)}\right) \quad (4.24)$$

where $(long_A, lat_A)$ is the longitude and latitude of Heathrow, $(long_B, lat_B)$ is the same for JFK, and R is the radius of the Earth, which is about 6378.137 km. Calculated in this way, the distance between Heathrow and JFK airports is 5554 km (3451 miles). Fortunately, you are unlikely ever to need the equation because suitable software can do the calculation for you.[11]

FIGURE 4.7 Showing the great-circle distance from Heathrow to JFK airports

[11]Such as the web interface at http://www.gpsvisualizer.com

4.9 THE SIGMA SIGN, Σ

The straight-line distance is a biased estimate of my actual distance to work. It will underestimate my journey because I must always travel a more circuitous route (either that or I run through people's homes and across a motorway). My actual route, as shown in Figure 4.6, is a a series of grid coordinates joined together, where each point arises from my phone receiving a GPS update about once every 4.5 seconds.

Table 4.2 shows the distance between the points for the first eight locations beginning after my house. In total, 741 distances were recorded during the journey. Adding all of them together gives a more accurate estimate of the total distance travelled: 10,736 metres.

TABLE 4.2 The first eight entries in the distance log

Location	1	2	3	4	5	6	7	8
Distance (m)	14.2	15.4	11.7	18.3	10.2	14.6	20.3	12.5

Denoting the first entry as x_1, the second x_2, the third x_3 (and so forth), the total distance is

$$d_{total} = x_1 + x_2 + x_3 + \cdots + x_{741} \tag{4.25}$$

where $\cdots$ means 'continue in the same way' and saves writing all the values in-between. The equation also could be written as

$$d_{total} = \sum_{i=1}^{n} x_i \tag{4.26}$$

where Σ (capital sigma) means 'add together' and the subscript i indicates we will add the values from the first to the last, where the last is referred to as the nth. For this example, $n = 741$. Equation (4.26) might also be written as $d_{total} = \Sigma x$, which means to add together all the values of x.

Back at the beginning of the chapter, the average – more specifically, the mean average – was described as what you get when you add together all the numbers and then divide their total by the number of numbers you had. Written using the sigma sign, it is

$$\bar{x} = \frac{\sum_{i=1}^{n} x_i}{n} \quad \text{or} \quad \bar{x} = \frac{\Sigma x}{n} \tag{4.27}$$

where $\bar{x}$, pronounced 'x bar', is the mean (see Chapter 5).

You may also see the symbol $\prod$. This is the multiplicative equivalent of Σ and used to find the product of some numbers. For example,

$$\prod_{i=1}^{10} a_i = a_1 \times a_2 \times a_3 \times a_4 \times a_5 \times a_6 \times a_7 \times a_8 \times a_9 \times a_{10} \qquad (4.28)$$

Another common expression is '!', which means factorial. A factorial is the product of a positive integer and all the other integers below it, down to 1. For example, $2! = 2 \times 1$, $3! = 3 \times 2 \times 1$, $4! = 4 \times 3 \times 2 \times 1$ and so forth. Factorials become very large, very quickly. For example, 10! is 3,628,800. If you throw n unbiased coins and want to know how many ways they can land k heads up, then it can be calculated as

$$\frac{n!}{k!(n-k)!}, \quad 0 \le k \le n, \qquad (4.29)$$

where $0 \le k \le n$ means that the number of heads must be greater than or equal to zero and less than or equal to the number of coins.

4.10 VECTORS AND MATRICES

We now turn to what I imagine will be the most off-putting part of the chapter, although it is not intended to be so. The days have gone when geographers spent long hours manipulating matrices by hand. I doubt those times are much missed. However, some knowledge of vector and matrix notation is important because it appears in analytical work and it's useful to know what you are looking at when you see it. In their book entitled *Matrices and Society*, Ian Bradley and Ronald Meek (2014 [1986]) demonstrate how matrix algebra can be used to show how power struggles in offices or committees develop, to predict how fast news or gossip will spread in a village, and to illuminate kinship structures in tribal societies. Matrices are widely used in statistics. For example, the position of the line of best fit shown in Figure 4.4 can be derived using the equation $\left[\mathbf{X}^{\mathsf{T}}\mathbf{X}\right]^{-1}\mathbf{X}^{\mathsf{T}}\mathbf{y}$, but this is only helpful if we have some understanding of what the mathematical gobbledygook means. Matrices are used in spatial statistics to represent the distance or the strength of connection between people and places.

Vectors

We begin with vectors. Looking back at Figure 4.4 and beginning at its leftmost side, the first three points on the graph are at (6.8, 33.9), (8.1, 34.7) and (8.4, 35.3). Imagine those data were recorded in a spreadsheet. Most likely they'd be entered as rows and columns: one row for each point, with one column for its X-value, and one column for its Y. What we'd then have would be like Table 4.3.

TABLE 4.3 An extract from the table of data used to produce Figure 4.4

Observation (neighbourhood)	% Population not of White ethnicity	% Population exhibiting deprivation
1	6.8	33.9
2	8.1	34.7
3	8.4	35.3
...	...	...

The same data can be represented as column vectors:

$$\mathbf{x} = \begin{bmatrix} 6.8 \\ 8.1 \\ 8.4 \\ ... \end{bmatrix}, \mathbf{y} = \begin{bmatrix} 33.9 \\ 34.7 \\ 35.3 \\ ... \end{bmatrix} \tag{4.30}$$

where $x_1 = 6.8$, $y_1 = 33.9$; $x_2 = 8.1$, $y_2 = 34.7$; $x_3 = 8.4$, $y_3 = 35.3$ and so forth. The vector is a collection of numbers.

Adding or subtracting vectors is straightforward, although it does require the same number of elements (values) in both.[12] For example,

$$\text{if } \mathbf{a} = \begin{bmatrix} 5 \\ 3 \\ 1 \end{bmatrix} \text{ and } \mathbf{b} = \begin{bmatrix} 4 \\ 2 \\ 8 \end{bmatrix}$$

$$\text{then } \mathbf{a} + \mathbf{b} = \begin{bmatrix} 5+4 \\ 3+2 \\ 1+8 \end{bmatrix} = \begin{bmatrix} 9 \\ 5 \\ 9 \end{bmatrix} \text{ and } \mathbf{a} - \mathbf{b} = \begin{bmatrix} 5-4 \\ 3-2 \\ 1-8 \end{bmatrix} = \begin{bmatrix} 1 \\ 1 \\ -7 \end{bmatrix} \tag{4.31}$$

Multiplying vectors by a single (scalar) value also is straightforward. For, example,

$$2\mathbf{a} = 2 \begin{bmatrix} 5 \\ 3 \\ 1 \end{bmatrix} = \begin{bmatrix} 10 \\ 6 \\ 2 \end{bmatrix}$$

---

[12]This is described as the vectors being conformable.

However, multiplying the vectors by each other is more complex: $\mathbf{ab}$ is <u>not</u> equal to $\begin{bmatrix} 5 \times 4 \\ 3 \times 2 \\ 1 \times 8 \end{bmatrix}$ but is equal to 34.

More strictly, it is $\mathbf{a}^\mathrm{T}\mathbf{b}$ that equals 34. That is because the number of *columns* in the first vector should be equal to the number of *rows* in the second when multiplying them together. Both $\mathbf{a}$ and $\mathbf{b}$ are 3 by 1 vectors, with 3 rows and 1 column. Therefore the number of columns in $\mathbf{a}$ is not equal to the number of rows in $\mathbf{b}$. However, if $\mathbf{a}$ is transposed – swapping its column with its rows – it produces a vector with 1 row and 3 columns:

$$\mathbf{a}^\mathrm{T} = \begin{bmatrix} 5 & 3 & 1 \end{bmatrix} \tag{4.32}$$

This transposed vector can then be multiplied by $\mathbf{b}$.

Multiplying vectors involves adding together their inner products, which in turn involves pairing column 1 in the first vector with row 1 in the second, column 2 in the first with row 2 in the second, and so forth in sequence:

$$\text{if} \quad \mathbf{a}^\mathrm{T} = \begin{bmatrix} a_1 & a_2 & a_3 & \cdots & a_n \end{bmatrix} \quad \text{and} \quad \mathbf{b} = \begin{bmatrix} b_1 \\ b_2 \\ b_3 \\ \cdots \\ b_n \end{bmatrix}$$

$$\text{then} \quad \mathbf{a}^\mathrm{T}\mathbf{b} = \left(a_1 b_1\right) + \left(a_2 b_2\right) + \left(a_3 b_3\right) + \cdots + \left(a_n b_n\right) = \sum_{i=1}^{n} a_i b_i \tag{4.33}$$

Therefore

$$\text{if} \quad \mathbf{a}^\mathrm{T} = \begin{bmatrix} 5 & 3 & 1 \end{bmatrix} \quad \text{and} \quad \mathbf{b} = \begin{bmatrix} 4 \\ 2 \\ 8 \end{bmatrix} \quad \text{then} \quad \mathbf{ab} = \left(5 \times 4\right) + \left(3 \times 2\right) + \left(1 \times 8\right) = 34$$

Table 4.4 provides a less abstract example. It refers again to my run to work and gives (a) the average speed to each point on the journey (in metres travelled per second) and (b) the time it took to reach the point from the previous one (in seconds). Multiplying the speed by the time taken gives the distance to each point (speed $\times$ time = distance). Adding those distances together gives the total distance travelled. Therefore, if $\mathbf{s}$ is a column vector of speeds and $\mathbf{t}$ is a column vector of times then $\mathbf{s}^\mathrm{T}\mathbf{t} = \sum_{i=1}^{n} s_i t_i$, which is 10,736 metres:

$$\text{if} \quad \mathbf{s} = \begin{bmatrix} 2.83 \\ 3.85 \\ 2.93 \\ \cdots \end{bmatrix} \quad \text{and} \quad \mathbf{t} = \begin{bmatrix} 5 \\ 4 \\ 4 \\ \cdots \end{bmatrix}$$

$$\text{then} \quad \mathbf{s}^{\mathrm{T}}\mathbf{t} = \left(2.83 \times 5 + 3.85 \times 4 + 2.93 \times 4 + \cdots\right) \tag{4.34}$$

TABLE 4.4 Average speed and time taken between points on my journey to work

Location	1	2	3	4	5	6	7	...	741
Average speed (metres per second)	2.83	3.85	2.93	3.66	2.54	3.64	4.05	...	3.13
Time taken (seconds)	5	4	4	5	4	4	5	...	4

Another example of vector multiplication that appears a lot in statistics is one of the form $\mathbf{x}^{\mathrm{T}}\mathbf{x}$. All this means is to take the values, square them, and sum those squares together. It calculates the sum of squares. For example, the column vector below gives the deviations around the line shown in Figures 4.1 and 4.2 and discussed at the beginning of the chapter:

$$\mathbf{x} = \begin{bmatrix} 1.34 \\ 0.78 \\ 1.22 \\ \cdots \\ -3.20 \end{bmatrix}$$

Calculating $\mathbf{x}^{\mathrm{T}}\mathbf{x}$ gives the sum of the squared deviations:

$$\mathbf{x}^{\mathrm{T}}\mathbf{x} = \left(1.34 \times 1.34\right) + \left(0.78 \times 0.78\right) + \left(1.22 \times 1.22\right) + \cdots + \left(-3.20 \times -3.20\right) \tag{4.35}$$
$$= 39.3$$

Matrices

Where vectors represent a single column or row of data, a matrix represents a more expansive collection. We might, for example, have

$$\mathbf{A} = \begin{bmatrix} 1 & -2 & 3 \\ 2 & -5 & 10 \\ -1 & 2 & -2 \end{bmatrix}, \mathbf{B} = \begin{bmatrix} 3 & 1 & 2 \\ 7 & 3 & 4 \\ 5 & 2 & 3 \end{bmatrix} \qquad (4.36)$$

where **A** and **B** are both 3 by 3 matrices. These are called square matrices because they have the same number of rows as they do columns. Alternatively, we might have

$$\mathbf{C} = \begin{bmatrix} 1 & 6 \\ 5 & 2 \\ 3 & 4 \end{bmatrix}, \mathbf{D} = \begin{bmatrix} 0 & -2 \\ 2 & -1 \\ 1 & 3 \end{bmatrix} \qquad (4.37)$$

where **C** and **D** are 3 by 2 matrices and not therefore square. As before, matrices can be added or subtracted if they have the same number of columns and rows:

$$\mathbf{A} + \mathbf{B} = \begin{bmatrix} 1+3 & -2+1 & 3+2 \\ 2+7 & -5+3 & 10+4 \\ -1+5 & 2+2 & -2+3 \end{bmatrix} = \begin{bmatrix} 4 & -1 & 5 \\ 9 & -2 & 14 \\ 4 & 4 & 1 \end{bmatrix}$$

$$\mathbf{A} - \mathbf{B} = \begin{bmatrix} 1-3 & -2-1 & 3-2 \\ 2-7 & -5-3 & 10-4 \\ -1-5 & 2-2 & -2-3 \end{bmatrix} = \begin{bmatrix} -2 & -3 & 1 \\ -5 & -8 & 6 \\ -6 & 0 & -5 \end{bmatrix} \qquad (4.38)$$

Like vectors, they easily are multiplied by a single (scalar) value:

$$0.5\mathbf{A} = \begin{bmatrix} 0.5 & -1 & 1.5 \\ 1 & -2.5 & 5 \\ -0.5 & 1 & -1 \end{bmatrix}, 2\mathbf{D} = \begin{bmatrix} 0 & -4 \\ 4 & -2 \\ 2 & 6 \end{bmatrix} \qquad (4.39)$$

However, multiplying the matrices by each other is more involved. As with vectors, the number of columns in the first matrix (**A**) needs to be equal to the number of rows in the second (**B**). It then gets more complicated. We begin with the first row of **A** and work through the columns of **B** in turn. So, the numbers in row 1 of **A** are multiplied by the numbers in column 1 of **B**, and added together. Then, still staying with row 1 of **A**, its numbers are multiplied by the numbers in column 2 of **B**, and added together. Next, row 1 of **A** with column 3 of **B**, and so on to the final column in **B**. Even then we are not finished because the whole process needs to be repeated for all the remaining rows in **A**. For instance,

$$\text{if } \mathbf{A} = \begin{bmatrix} a_{11} & a_{12} & a_{13} \\ a_{21} & a_{22} & a_{23} \\ a_{31} & a_{32} & a_{33} \end{bmatrix} \text{ and } \mathbf{B} = \begin{bmatrix} b_{11} & b_{12} & b_{13} \\ b_{21} & b_{22} & b_{23} \\ b_{31} & b_{32} & b_{33} \end{bmatrix}$$

then $\mathbf{AB}$

$$= \begin{bmatrix} a_{11}b_{11} + a_{12}b_{21} + a_{13}b_{31} & a_{11}b_{12} + a_{12}b_{22} + a_{13}b_{32} & a_{11}b_{13} + a_{12}b_{23} + a_{13}b_{33} \\ a_{21}b_{11} + a_{22}b_{21} + a_{23}b_{31} & a_{21}b_{12} + a_{22}b_{22} + a_{23}b_{32} & a_{21}b_{13} + a_{22}b_{23} + a_{23}b_{33} \\ a_{31}b_{11} + a_{32}b_{21} + a_{33}b_{31} & a_{31}b_{12} + a_{32}b_{22} + a_{33}b_{32} & a_{31}b_{13} + a_{32}b_{23} + a_{33}b_{33} \end{bmatrix}$$

i.e.

$$\begin{bmatrix} \text{'A row 1 with B column 1'} & \text{'A row 1 with B column 2'} & \text{'A row 1 with B column 3'} \\ \text{'A row 2 with B column 1'} & \text{'A row 2 with B column 2'} & \text{'A row 2 with B column 3'} \\ \text{'A row 3 with B column 1'} & \text{'A row 3 with B column 2'} & \text{'A row 3 with B column 3'} \end{bmatrix} \quad (4.40)$$

This isn't pretty, so let's take a simpler example. Imagine that matrix $\mathbf{F}$ represents the flows of pupils from three neighbourhoods into two schools, where the rows are the neighbourhoods and the columns are the schools. Further imagine that matrix $\mathbf{D}$ represents the distance (in kilometres) from the neighbourhoods to the schools:

$$\mathbf{F} = \begin{bmatrix} 100 & 80 \\ 200 & 50 \\ 200 & 100 \end{bmatrix}, \mathbf{D} = \begin{bmatrix} 2 & 2 \\ 1 & 4 \\ 6 & 3 \end{bmatrix}$$

Because the columns in $\mathbf{F}$ are fewer than the number of rows in $\mathbf{D}$, we can't immediately multiply $\mathbf{F}$ and $\mathbf{D}$. However, we can transpose $\mathbf{F}$ and proceed from there:

$$\mathbf{F}^T = \begin{bmatrix} 100 & 200 & 200 \\ 80 & 50 & 100 \end{bmatrix}, \mathbf{D} = \begin{bmatrix} 2 & 2 \\ 1 & 4 \\ 6 & 3 \end{bmatrix}$$

$$\mathbf{F}^T\mathbf{D} = \begin{bmatrix} 100 \times 2 + 200 \times 1 + 200 \times 6 & 100 \times 2 + 200 \times 4 + 200 \times 3 \\ 80 \times 2 + 50 \times 1 + 100 \times 6 & 80 \times 2 + 50 \times 4 + 100 \times 3 \end{bmatrix}$$

$$\mathbf{F}^T\mathbf{D} = \begin{bmatrix} 1600 & 1600 \\ 810 & 660 \end{bmatrix} \quad (4.41)$$

In the final matrix, the top left entry is the total distance travelled by pupils from the neighbourhoods to the first school, and the bottom right is the distance travelled by pupils to the second. Together these distances are less than if all the pupils switched school (shown by the remaining values in the matrix). In the sense of minimising travel, the current arrangements are better than if everyone switched schools.

There is much more that could be said about matrices. However, we will limit ourselves to three observations. First, matrix multiplication is not commutative: there is no guarantee that $\mathbf{AB}$ (or $\mathbf{A}^T\mathbf{B}$) will be equal to $\mathbf{BA}$ (or $\mathbf{B}^T\mathbf{A}$). Because of this, and for clarity, a distinction can be made between pre-multiplying a matrix ($\mathbf{AB}$ is $\mathbf{B}$ pre-multiplied by $\mathbf{A}$) and post-multiplying ($\mathbf{BA}$ is $\mathbf{B}$ post-multiplied by $\mathbf{A}$).

Second, there has been no mention of division. That is because it makes little sense to talk about matrix division. However, by analogy with the fact that $\dfrac{a}{b} = a \times \dfrac{1}{b} = ab^{-1}$ (where a and b are two 'ordinary' numbers and b^{-1} is the inverse of b), we can talk about multiplying one matrix by the inverse of the other.

When a number is multiplied by its inverse, the answer is 1 (e.g. $2 \times \dfrac{1}{2} = 1$). The equivalent condition for a matrix is the identity matrix, $\mathbf{I}$, which is a square matrix comprised of zeros, except along the leading diagonal, which contains 1s. For example, $\mathbf{I}$ could be $\begin{bmatrix} 1 & 0 \\ 0 & 1 \end{bmatrix}$ or $\begin{bmatrix} 1 & 0 & 0 \\ 0 & 1 & 0 \\ 0 & 0 & 1 \end{bmatrix}$ or $\begin{bmatrix} 1 & 0 & 0 & 0 \\ 0 & 1 & 0 & 0 \\ 0 & 0 & 1 & 0 \\ 0 & 0 & 0 & 1 \end{bmatrix}$ and so forth.

In principle $\mathbf{AA}^{-1} = \mathbf{I}$ and $\mathbf{BB}^{-1} = \mathbf{I}$, etc. However, not every matrix has an inverse. Only square matrices do, but not every one of them. Without going into detail about how it was calculated we can show that matrix $\mathbf{A}$ in equation (4.36) does have an inverse:

$$\mathbf{A} = \begin{bmatrix} 1 & -2 & 3 \\ 2 & -5 & 10 \\ -1 & 2 & -2 \end{bmatrix}, \mathbf{A}^{-1} = \begin{bmatrix} 10 & -2 & 5 \\ 6 & -1 & 4 \\ 1 & 0 & 1 \end{bmatrix}, \mathbf{AA}^{-1} = \begin{bmatrix} 1 & 0 & 0 \\ 0 & 1 & 0 \\ 0 & 0 & 1 \end{bmatrix} \quad (4.42)$$

However, matrix $\mathbf{B}$ in the same equation does not. Using my statistical software to invert that matrix produces an error message: 'system is computationally singular'. The problem is that although $\mathbf{B}$ has three columns, one of them is essentially duplicating the information found in the other two. As we go down the rows of matrix $\mathbf{B}$, we find that the first element is equal to the sum of the second and third:

in row 1, $3 = 1 + 2$; in row 2, $7 = 3 + 4$; in row 3, $5 = 2 + 3$. The consequence is that there is less information than is needed for the inverse to be determined.

Third, in geographical applications, a spatial weights matrix is used to represent the connections between places. Consider the map in Figure 4.8, which shows the Australian states and territories. A map displays spatial relationships. One of those is which places are neighbours. A matrix can be used to encode that information and the matrix, **W**, in equation (4.43) does so, showing which of the states and territories – Western Australia (WA), Northern Territory (NT), South Australia (SA), Queensland (QL), New South Wales (NS), Victoria (VI), Tasmania (TZ) and Australian Capital Territory (AC) – share a border. A value of 1 is given to contiguous (bordering) locations (Western Australia and Northern Territory, for example), and zero otherwise. The matrix is both square and symmetrical (if Western Australia shares a boundary with Northern Territory then Northern Territory must also share a boundary with Western Australia). It follows convention in saying that a state is not contiguous with itself.

FIGURE 4.8 The Australian states and territories. The map shows which locations share a border and is used to generate matrix, **W**, in equation (4.43)

$$
\mathbf{W} = \begin{array}{c|cccccccc}
 & \text{(WA)} & \text{(NT)} & \text{(SA)} & \text{(QL)} & \text{(NS)} & \text{(VI)} & \text{(TZ)} & \text{(AC)} \\
\text{(WA)} & 0 & 1 & 1 & 0 & 0 & 0 & 0 & 0 \\
\text{(NT)} & 1 & 0 & 1 & 1 & 0 & 0 & 0 & 0 \\
\text{(SA)} & 1 & 1 & 0 & 1 & 1 & 1 & 0 & 0 \\
\text{(QL)} & 0 & 1 & 1 & 0 & 1 & 0 & 0 & 0 \\
\text{(NS)} & 0 & 0 & 1 & 1 & 0 & 1 & 0 & 1 \\
\text{(VI)} & 0 & 0 & 1 & 0 & 1 & 0 & 0 & 0 \\
\text{(TZ)} & 0 & 0 & 0 & 0 & 0 & 0 & 0 & 0 \\
\text{(AC)} & 0 & 0 & 0 & 0 & 1 & 0 & 0 & 0 \\
\end{array}
\qquad (4.43)
$$

Summing along the rows of the matrix (or, alternatively, down the columns) we arrive at the following information: (WA) 2; (NT) 3; (SA) 5; (QL) 3; (NS) 4; (VI) 2; (TZ) 0; (AC) 1. The sum tells us that Southern Australia has the most contiguous neighbours, whilst Tasmania has none.

Multiplying the matrix by itself creates a powered matrix:

$$
\mathbf{W}^2 = \mathbf{WW} = \begin{array}{c|cccccccc}
 & \text{(WA)} & \text{(NT)} & \text{(SA)} & \text{(QL)} & \text{(NS)} & \text{(VI)} & \text{(TZ)} & \text{(AC)} \\
\text{(WA)} & 2 & 1 & 1 & 2 & 1 & 1 & 0 & 0 \\
\text{(NT)} & 1 & 3 & 2 & 1 & 2 & 1 & 0 & 0 \\
\text{(SA)} & 1 & 2 & 5 & 2 & 2 & 1 & 0 & 1 \\
\text{(QL)} & 2 & 1 & 2 & 3 & 1 & 2 & 0 & 1 \\
\text{(NS)} & 1 & 2 & 2 & 1 & 4 & 1 & 0 & 0 \\
\text{(VI)} & 1 & 1 & 1 & 2 & 1 & 2 & 0 & 1 \\
\text{(TZ)} & 0 & 0 & 0 & 0 & 0 & 0 & 0 & 0 \\
\text{(AC)} & 0 & 0 & 1 & 1 & 0 & 1 & 0 & 1 \\
\end{array}
\qquad (4.44)
$$

The information it contains is the number of ways of getting from one location to another in two steps, where a step is a movement across a boundary. For example, Queensland can be reached from Western Australia in two steps, via either Northern Territory or South Australia. A two-step, return journey from and to Western Australia can be made via the same. The powered matrix, $\mathbf{W}^3$, yields the number of ways a three-step journey can be made, and so forth.

Abler et al. (1972: Chapter 8) describe a process by which the addition of successive powers of $\mathbf{W}$ (i.e. $\mathbf{W}$, $\mathbf{W} + \mathbf{W}^2$, $\mathbf{W} + \mathbf{W}^2 + \mathbf{W}^3$, …) is used to create a new matrix, the process stopping when that matrix contains no zero entries. That will never happen for our example because Tasmania is a disconnected island. However, we can stop when all the other entries reach zero, which is at the third power:

$$
\mathbf{W} + \mathbf{W}^2 + \mathbf{W}^3 =
\begin{array}{c}
\begin{array}{cccccccc}
\quad & (WA) & (NT) & (SA) & (QL) & (NS) & (VI) & (TZ) & (AC)
\end{array} \\
\begin{array}{c}
(WA) \\ (NT) \\ (SA) \\ (QL) \\ (NS) \\ (VI) \\ (TZ) \\ (AC)
\end{array}
\left[
\begin{array}{cccccccc}
4 & 7 & 9 & 5 & 5 & 3 & 0 & 1 \\
7 & 7 & 11 & 9 & 6 & 5 & 0 & 2 \\
9 & 11 & 13 & 12 & 12 & 9 & 0 & 3 \\
5 & 9 & 12 & 7 & 10 & 5 & 0 & 2 \\
5 & 6 & 12 & 10 & 8 & 8 & 0 & 5 \\
3 & 5 & 9 & 5 & 8 & 4 & 0 & 2 \\
0 & 0 & 0 & 0 & 0 & 0 & 0 & 0 \\
1 & 2 & 3 & 2 & 5 & 2 & 0 & 1
\end{array}
\right]
\end{array}
\quad (4.45)
$$

Summing along the rows gives: (WA) 34; (NT) 47; (SA) 69; (QL) 50; (NS) 54; (VI) 36; (TZ) 0; (AC) 16. It tells us that in locational terms, South Australia is the most connected state and (excluding Tasmania), Australian Capital Territory the least. This was always obvious from the map. However, the principle applies to situations when it is not so self-evident: the most accessible capital by flight paths or the most accessible station on the New York Subway, for example.

4.11 CONCLUSION

We have travelled a long way in this chapter, from the simplicity of knowing that 2×5 equals 5×2 to the complexity of understanding that **AB** may not be equal to **BA**, and that not all matrices can be inverted. As we reach the end, you would probably agree that some level of numeracy is important in almost any university-level subject, yet you may also wonder if everything that has been covered here is necessary to study geography. The answer I would give is yes, -ish, sort of. To say this is not to deny that you could probably avoid using maths in your studies. Nor is it to suggest that maths is the only skill essential for your learning and understanding; reading, written communication, clear reasoning, developing an argument based on evidence, being able to see a different point of view, listening, being sensitive to difference − these are all important, and the list easily could be extended. Yet, maths remains important in the science, social sciences and in the humanities too. The point is not that you have to be a mathematician to practise geography, but that it is helpful to have a little mathematical knowledge to gain a greater appreciation of the kinds of studies that geographers undertake and − even better − to join in.

5

DESCRIPTIVE AND INFERENTIAL STATISTICS

5.1 INTRODUCTION

Chapters 3 and 4 have introduced some statistical and mathematical foundations upon which this chapter now builds. It looks at common ways to summarise a set of data, including determining their average and variation. These are often referred to as descriptive statistics, in that they describe the centre and spread of the data and allow comparisons of different variables to be made. An extension to this idea is to calculate a statistic for different parts of a study region and see how the statistic varies across the study region – a geographically minded approach that we touch upon briefly in this chapter and return to again in Chapter 11.

This chapter also introduces the normal (or Gaussian) curve as a distribution that appears a lot in statistics. In fact, many data are not normal but skewed, meaning they have more high values than they do low ones or vice versa. Although it may be possible to transform skewed data to produce a normal distribution, often the assumption of normality lies not in the shape of the data but in the sampling distribution that would arise if it were possible to sample from a population over and over again. The distinction between a sample and a population is the difference between incomplete data about what is being measured and the complete set of data that would emerge were it possible to enumerate the population in its entirety. The interest in a sample can extend beyond simply describing the data in their own right to using them to shed light on the statistical properties of the population from which they are drawn. Using a sample to solicit information about the population is known as statistical inference and leads to ideas such as confidence intervals, which end this chapter.

5.2 MEASURES OF CENTRAL TENDENCY (AVERAGES)

The World Values Survey is, in fact, a collection of surveys undertaken in almost 100 countries across the world. It asks questions about the beliefs and values of their populations. Its website reports that 'thousands of political scientists, sociologists, social psychologists, anthropologists and economists have used these data to analyse such topics as economic development, democratization, religion, gender equality, social capital, and subjective well-being'.[1] Government officials and organisations such as the World Bank also use it.

The questionnaire can be seen online where there are tools to view, analyse and download the data. One of the many questions asked is the respondents' support for the statement that an essential characteristic of democracy is for the state to make people's incomes equal. Specifically, each respondent is asked to give a score from 1 to 10, where 1 means they do not see this as the state's role at all whereas 10 means they view state involvement as an essential characteristic. For the 2010–14 data, 82,297 people from 60 countries answered the question. The average score was 5.96, suggesting a slight leaning towards positively viewing state-based policies for income equality.

What is meant by average has been largely undefined to this point. The classic definition is the one suggested in Chapter 4: the result obtained by adding a set of numbers together and dividing their total by the amount of numbers in the set. This is what most people mean when they talk of an average. In notation,

$$\bar{x} = \frac{\sum_{i=1}^{n} x_i}{n} \quad \text{or} \quad \bar{x} = \frac{\sum x}{n} \tag{5.1}$$

where $\bar{x}$ ('x bar') is the average, x is a value in the set, $\sum$ means sum together and n is the number of values in the set. Each value may be referred to as an 'observation', since each is a measurement of something that has been observed of a person, place or some other social or physical feature. Therefore x denotes an observation and n is the number of observations. Refer back to Section 4.9 if you are unclear about the notation.

What equation (5.1) defines is the mean average. It is the number at the centre of the data in the sense that if we sum the differences between each observation and the mean, the numbers above the mean will balance out the values below, giving an overall sum of zero:

$$\sum_{i=1}^{n} (x_i - \bar{x}) = 0, \quad \text{i.e.} \quad (x_1 - \bar{x}) + (x_2 - \bar{x}) + (x_3 - \bar{x}) + \cdots + (x_n - \bar{x}) = 0 \tag{5.2}$$

The mean is the numeric equivalent of a pivot point where the moment (the force) on one side of the pivot is balanced by the moment on the other side, as in Figure 5.1. However, it would be mistaken to assume that it is also the middle value, in the sense that if all the data were ordered from lowest to highest the mean

---

[1]http://www.worldvaluessurvey.org/WVSContents.jsp

would necessarily be halfway. It need not and usually will not be. The value that is at the midpoint is called the 'median'. If n data values are sorted from lowest to highest, and n is an odd number, then the median is the $\frac{1}{2}(n+1)$th of them. For example, it is the value corresponding to the 41,149th of the 82,297 sorted responses to the survey question. In this instance, it is equal to 6. A problem arises if n is an even number – halfway from 1 to 10 is the 5.5th value, but that doesn't exist. In such instances it will be necessary to pick one or other of the two values either side of the middle (the 5th or 6th where $n = 10$), or take them both and calculate their mean.

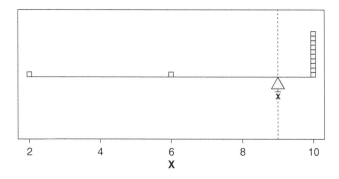

FIGURE 5.1 The mean average is like a pivot point balancing the lower and higher numbers

The median is considered to be a robust statistic in a way that the mean is not. This is because the median is less affected by outliers, which are atypically high or low values. According to the American Community Survey, the mean average household income was $73,487 in 2013.[2] The problem with this estimate is that the average is inflated by a relatively small number of very high-earning house-holds. The fact that only 35 per cent of households earned more than $75,000 suggests the mean is not in the middle of the earnings distribution but pulled up by the higher values. Therefore the median income is a better indicator of what the average family is earning. It was estimated to be $53,046, which is 28 per cent less than the mean.[3]

A third, more occasionally used, average is the value most frequently occurring in the data, called the modal value or 'mode'. Figure 5.2 plots how often each of the possible responses was chosen in answer to the World Values Survey question

---

[2]http://factfinder.census.gov/
[3]Sometimes a trimmed mean is used. This trims off a given proportion of the highest and lowest values, giving the potential to remove outliers.

about the importance of income equality as a characteristic of democracy. The mode is 10, suggesting the average respondent sees income equality as an essential characteristic of democracy. This would be encouraging to campaigners for income equality, but looking at Figure 5.2 shows this average to be deceptive: yes, 10 is the most frequent answer, but 5 and the opposite extreme of 1 are commonly occurring too. In fact, none of the averages captures the diversity of opinion well. They are not intended to. They are measures of central tendency, not of variation around it. Nevertheless, the mode appears particularly susceptible to giving undue attention to a result that, although most common, is not actually that common. For the survey, only 17 per cent of respondents gave the modal value of 10.

The mode is also limited in when it can be applied. For the survey question the permissible answers are limited to a small number of distinct (discrete) values. However, many measurements and observations are not constrained in this way. Instead, they could take on any value, within a minimum and maximum range, at least, and to the precision of the measurement device. This is the difference between discrete and continuous data. It is unwise to calculate the mode for continuous data, because there are so many possible answers that any one will appear rarely. The mode may nevertheless be suitable for data consisting of nominal categories. Nominal means categorical (sometimes called 'qualitative') data when there is no reason to rank one category as numerically higher than another. For example, if all the readers of this book were asked to select their favourite

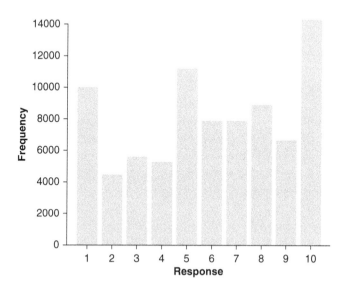

FIGURE 5.2 Frequency of responses to the World Values Survey question about whether the state making incomes equal is an essential characteristic of democracy (1 = not at all essential; 10 = definitely essential) (source: author's own calculations using data from www.worldvaluessurvey.org)

geographer from a list of Peter Haggett, Richard Hartshorne, Ron Johnston, Denis Cosgrove, Nigel Thrift, Robert Mayhew, Joe Lobley, Sarah Whatmore, Mei-Po Kwan and Richard Dastardly, calculating the mean or median result would be impossible because names are not numbers. The mode can be determined. It is the most popular answer.

5.3 WEIGHTED AVERAGES

Table 5.1 shows the estimated percentage of the population living below the poverty line for various counties in California. Using equation (5.1) to average across these values yields the mean average poverty rate for the counties. It is 17.3 per cent. What it doesn't provide is an accurate estimate of the percentage of the counties' total population that is living below the poverty line. That is 16.8 per cent.

TABLE 5.1 Percentage of the population estimated to be living below the poverty line in selected Californian counties, 2013 (source: American Community Survey/Census Bureau's Population Estimates Program)

County	%	County	%	County	%
Alameda County	13.0	Madera County	24.1	San Mateo County	8.0
Amador County	15.9	Marin County	8.3	Santa Barbara County	15.9
Butte County	22.0	Mendocino County	21.0	Santa Clara County	10.6
Calaveras County	12.8	Merced County	26.0	Santa Cruz County	14.7
Colusa County	13.4	Monterey County	17.5	Shasta County	19.0
Contra Costa County	11.3	Napa County	11.1	Siskiyou County	22.5
Del Norte County	22.2	Nevada County	13.0	Solano County	13.9
El Dorado County	10.3	Orange County	13.1	Sonoma County	12.2
Fresno County	27.4	Placer County	9.0	Stanislaus County	21.9
Glenn County	20.7	Riverside County	17.2	Sutter County	17.3
Humboldt County	22.3	Sacramento County	18.7	Tehama County	19.0
Imperial County	24.2	San Benito County	11.0	Tulare County	28.4
Kern County	23.6	San Bernardino County	19.7	Tuolumne County	15.2
Kings County	21.4	San Diego County	15.1	Ventura County	11.6
Lake County	25.7	San Francisco County	14.2	Yolo County	19.4
Lassen County	18.8	San Joaquin County	18.7	Yuba County	21.1
Los Angeles County	18.8	San Luis Obispo County	15.0		

The discrepancy arises because they are subtly different measurements. The first takes the set of poverty rates per county and averages it. The second calculates the average poverty rate for the total population across all the counties. To obtain this second answer from the data available, the calculation needs to take into consideration that some counties have larger populations than others (a situation that the first measure ignores). A weighted average may be used where the weight, w, is the population count per county:

$$\bar{x} = \frac{\sum_{i=1}^{n} w_i x_i}{\sum_{i=1}^{n} w_i} \quad \text{or} \quad \bar{x} = \frac{\sum wx}{\sum w} \tag{5.3}$$

Each observation is now multiplied by its weight of population and the division is by the sum of weights instead of the number of observations.[4]

For another use of weighting, imagine we are interested in how support for income equality as a characteristic of democracy is viewed in Zimbabwe compared to Cyprus. The mean score for Zimbabwean respondents is 4.12, and for Cypriots it is 5.92. However, these averages ignore the survey weights that are intended to correct for the fact that survey respondents are not a demographically representative sample of the populations from which they are drawn. In this instance, calculating the weighted mean for the countries reduces the difference between them. It is now 4.28 for Zimbabwe and 5.79 for Cyprus.

Weighted averages are not limited to the mean. It is possible to calculate other weighted statistics too. Whereas the median is the value halfway along a numerically ordered set of values, the weighted median is the value halfway along the sum of their weights. Whereas the mode is the most frequently occurring value, the weighted mode is the value for which the sum of weights is greatest.

A geographical variation on weighting is to calculate a local average based on the values around a location. Suppose we want to calculate the average of the poverty rates in and around Orange County, as shown in Figure 5.3. The weights are a spatial weight, as discussed in Chapter 4, set to 1 if a county borders Orange County and 0 otherwise. In this instance, a weight of 1 is also assigned to Orange

---

[4]If this isn't clear, then imagine you are interested in the crime rate for the United States but for some bizarre reason have only two pieces of information: (1) the crime rate for Wyoming and (2) the combined crime rate for all other states, including Washington, DC, but excluding Wyoming. It ought to be obvious that you cannot just take the two values and average them. To do so would treat the one state, Wyoming, as equivalent to 50 others. Instead, you would need to allow for the fact that Wyoming contains less than 0.2 per cent of the US population and weight accordingly.

County itself so it is included in the local average. Going around the map from left to right, the calculation for the area shown in Figure 5.3 is

$$\bar{x}_{(u,v)} = \frac{\sum wx}{\sum w}$$

$$= \frac{(0 \times 15.0) + (0 \times 15.9) + \cdots + (1 \times 18.8) + (1 \times 13.1) + (1 \times 19.7) + \cdots + (0 \times 24.2)}{0 + 0 + \cdots + 1 + 1 + 1 + \cdots + 0}$$ (5.4)

where the subscript (u, v) indicates that the mean is a local statistic calculated for a specific part of the map. Note that all we are doing here is calculating the average per-county rate for Orange County and its four contiguous neighbours. If what we want is the combined poverty rate for Orange County and its neighbours then we need to weight geographically and by population size. That simply requires replacing all the non-zero weights with the population count for those counties. This sort of geographical approach to statistics is considered further in Chapter 11.

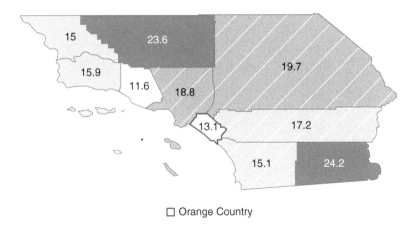

□ Orange Country

FIGURE 5.3 Percentage of the population estimated to be living below the poverty line in southern Californian counties, 2013. Orange County and its contiguous neighbours are highlighted

5.4 MEASURES OF VARIATION AND SPREAD

An average indicates the centre of some data but says nothing about how greatly individual observations deviate from that average. A set of data with greater over-all deviation is said to be more variable – to contain greater variation – than a set of data with less. Knowledge of this variation is important because it provides information about the range of outcomes that can be expected for whatever has been measured. When assessing flood risk the average water level is useful to know, but what is more important is how variable the levels in the water are: are they steady throughout the year or can they rise sharply, putting people at risk from flooding? Two rivers can have the same average but it is the one with greater variation that is likely to pose the greater danger.

The interquartile range (IQR)

Consider that in the academic year, 2011–12, the straight-line distance between a London pupil's home and their secondary (high) school had a mean average of 2.45 km. Did everyone travel 2.45 km to school? Of course not! Even ignoring that pupils must travel a more circuitous route, the average conceals variation. For one pupil the distance was 20.0 km. For some others it was zero.[5] These are the range of the data from the minimum to the maximum value. However, they are also the extremes. It is by no means common for a pupil to travel 20 km to school (which is over 12 miles). In fact, half the pupils travelled between 0.921 and 3.28 km, thresholds that are a little further than zero and much less than 20. I am highlighting these particular distances because they delimit the middle half of the data, as Figure 5.4 shows. In statistical language, these values are the interquartile range

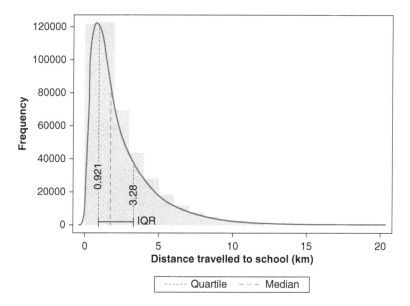

FIGURE 5.4 Distances travelled from home to school in London in 2011-12 (source: author's calculations based on National Pupil Database data)

[5]Not because the pupils lived in the school but because the measurement is from the postcode of their home to the postcode of the school, and those postcodes could be the same.

(IQR). Interquartile means 'between quarters'. The IQR is obtained by sorting the data set from lowest to highest and then working from one quarter to three quarters along it. The first quartile is 0.921 km. The third quartile is 3.28 km. Between the two is the second quartile, which is halfway and better known as the median.

Like the median, the IQR is regarded as a robust statistic. By considering only the 'mid-spread' of the data, it is not susceptible to outliers that are further to the extremes. That's not an argument to ignore the outliers, though. The typical height of the River Frome flowing into Bristol is between 0.31 and 1.12 metres, but that didn't stop it from reaching 2.42 metres on Christmas Eve 2013 and bursting its banks as a consequence.[6] It's just to say that the IQR focuses on the more common occurrences.

The minimum, maximum and median values are often combined with the IQR and the mean and reported in a six-number summary, as in Table 5.2, which reports the distances from home to school for pupils in London and also in the rest of South East England. The same information, excluding the mean, can be displayed pictorially using a box plot. Box plots give a visual impression of the shape of the data, their centre and spread. They also are useful for comparing data sets.

Like Table 5.2, the box plots in Figure 5.5 summarise the distance to school data for London and for the rest of South East England. Each IQR is represented by a rectangle, the median by the thicker line within it. The dashed lines extending out from the box are called the whiskers. In this example they extend up to 2.5 times the IQR from the edge of the box, or to the minimum or maximum value if that comes sooner. The default for the software is 1.5 times the IQR. Because it can be changed, it is worth checking what value your software is using. In any case, the purpose of the whiskers is to highlight values beyond the box. These could be outliers. There are lots of these for the schools data because it is such a large data set (436,579 observations for London, 492,954 for the rest of the South East). They are all found beyond the right-hand whisker, which tells us something more about the shape of the data. Specifically, it is skewed, meaning the data are not distributed symmetrically around their mean. The asymmetry can be seen in Figure 5.4.

TABLE 5.2 A six-number summary of the distance to school data, comparing London and the rest of South East England

	Min	Q1	Median	Mean	Q3	Max
London	0	0.921	1.73	2.48	3.28	20.0
Rest of SE	0	1.02	1.98	3.35	4.40	20.0

[6]http://apps.environment-agency.gov.uk/river-and-sea-levels/

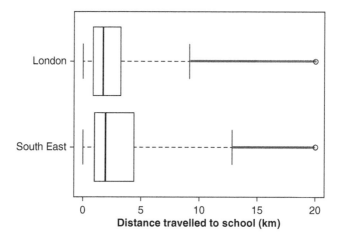

FIGURE 5.5 Box plots indicating and comparing the distances travelled from home to school in London and the rest of South East England. The asymmetry (skew) of both data sets is evident, as is the greater spread of distances in the South East. In each case the width of the box is the interquartile range

Sometimes potential outliers are worth checking for error, treating as special cases or even dropping from the data. This might make sense if, for example, there were one or two observations clearly different from the rest. However, this isn't the case for the pupil data. The longer distances may not be typical but they are not rare enough to warrant omission.[7] Comparing the box plots, the median distance travelled from home to school is greater in the South East than in London, and the spread of distances is greater in the South East too, with more pupils travelling further. This is not at all surprising: the density of housing and of schools is lower in the South East, forcing greater travel.

Other quantiles

Although the interquartile range is widely used, there is no particular reason to restrict attention to the middle half of the data. We could, for example, take the middle 80 per cent, which is found between the first and the ninth decile. This gives a range from 0.499 to 5.37 km for the distances to school in London. Alternatively, we might choose to consider all but the most extreme 2 per cent of values, ranging from the first to the 99th percentile. This, for the same data, is from 0.169 to 11.6 km. Since these values include almost all the data, they emphasise

---

[7]The box plots do reveal something else, however: for both regions the maximum distance travelled is 20 km. That seems a bit suspicious. Why should they be the same? It is because pupils travelling longer distances have already been removed from the data.

how unusual a value of 0 or 20 km is. Tertiles, quartiles, quintiles, deciles and percentiles are all ways of dividing the data into three, four, five, ten or 100 groups, respectively, where each group contains one third, one tenth, one quarter, one fifth, one tenth or one hundredth of the data. 'Quantiles' is the generic term for these and other ways of splitting the data into equally sized groups (or as equal as possible given the number of observations and the possibility of tied values).

The variance and standard deviation

Two other measures of spread are the variance and the standard deviation. Both are used to summarise how much the data deviate from the mean. You may sometimes see the variance calculated as

$$s^2 = \frac{\sum (x - \bar{x})^2}{n} \tag{5.5}$$

where s^2 denotes the variance. Within the top part of the equation, $(x - \bar{x})$ acts to mean-centre the data so that values greater than the mean are positive, values less than the mean are negative, and any value equalling the mean becomes zero. Squaring and adding those values gives the sum of the squared deviations around the mean, $\sum (x - \bar{x})^2$. Using vector notation, if $\mathbf{x}$ is a vector of mean-centred values then the sum of squares is $\mathbf{x}^T\mathbf{x}$. Dividing by n calculates an average – the average squared deviation. This is basically what the variance is. However, it is more common to see the variance calculated as

$$s^2 = \frac{\sum (x - \bar{x})^2}{n - 1} \tag{5.6}$$

and its close cousin, the standard deviation, calculated as

$$s = \sqrt{\frac{\sum (x - \bar{x})^2}{n - 1}} \tag{5.7}$$

Working back from equation (5.7) to (5.6), it is easy to see that the variance is the standard deviation squared. That is why the variance is denoted s^2, where s is the standard deviation. Less clear is why the division is by $n - 1$. I could say it is due to convention and leave it at that. A fuller answer is that we are dividing by the degrees of freedom instead of by the number of observations. The degrees of freedom are the number of independent values that go into the calculation and which could, in principle, be varied subject to one or more constraints upon them. In this case the constraint is the value of the mean. Holding that constant, then with a set of n numbers, up to $n - 1$ can be altered before the final value is fixed by the need for everything to balance up to give the required mean. The degrees of freedom can be defined as the sample size minus the number of parameters estimated from the data (Crawley, 2005). The mean is the only parameter that appears in equations (5.6) and (5.7), hence there are $n - 1$ degrees of freedom.[8]

5.5 Z-VALUES AND THE NORMAL DISTRIBUTION

Unlike the IQR, neither the variance nor the standard deviation is considered robust to outliers. Because the deviation from the mean is squared, unusually high or low values can have considerable influence upon the amount of variation observed in the data. Nevertheless, the variance and standard deviation are used widely in statistics. Applications include calculating standardised scores, also known as z-values. These are calculated by mean–centring each value and dividing by the standard deviation of the data:

$$z_x = \frac{x - \bar{x}}{s} \tag{5.8}$$

The result is to transform the original values into units of standard deviation around the mean. One advantage of this is to provide a way of identifying some of the more unusual values in a data set, doing so with consideration to their centre and spread. Assume there are two data sets, each with a mean of 100 but one containing a value of 105, the other a value of 150. Intuitively the second value seems the more unusual because it is further from its mean. However, that intuition needs to allow for how noisy and varied the data are. If the first data set has a standard deviation of 2 and the second has a standard deviation of 50, then it is the value of 105 that is the more unusual because it is at 2.5 standard deviations above its mean $\left(z = 2.5 = (105 - 100) / 2 \right)$, whereas the value of 150 is only one standard deviation above its $\left(z = 1 = (150 - 100) / 50 \right)$. The logic is that large deviations from the mean are not surprising in something that itself varies a lot, whereas they are surprising in something that only varies a little. The z-values provide a framework for judging how far a value is from the centre with respect to the spread of the data.

Often it is z-values of magnitude greater than 1.96 or 2.58 that attract the most interest; that is, values greater than +1.96 or +2.58, or less than −1.96 or −2.58. The reason for this lies in the properties of the curve shown in Figure 5.6. The

---

[8]A further answer is that the variance of a sample of data is usually the best estimate we have of the variance of the population from which the sample was drawn but that estimate is biased (too low) if using equation (5.5) instead of (5.6).

curve is sometimes known as the bell-shaped curve, or more formally as the normal or Gaussian distribution. If a data set has this shape when plotted using a histogram, for example (see Chapter 7), then 95 per cent of the values will lie within 1.96 standard deviations either side of the mean and 99 per cent will lie within 2.58 standard deviations. That means if we select a value at random from the data set, the probability it will be *more* than 1.96 standard deviations from the mean is only 0.05 (the equivalent of 5 per cent but expressed on a probability scale) and the probability it will be more than 2.58 standard deviations away is 0.01. Written algebraically, $P(|z| > 1.96) = 0.05$ and $P(|z| > 2.58) = 0.01$ if the data are normally distributed.

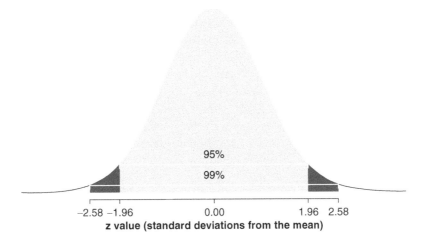

FIGURE 5.6 For normally distributed data, 95 per cent of the observations are within 1.96 standard deviations of the mean, and 99 per cent are within 2.58 standard deviations

The normal curve is actually a family of curves with different means and standard deviations, but each with the same mathematical formula describing their shape. Whilst the detail of that formula need not detain us, it is instructive to consider the process that generates a normal distribution.[9] It occurs when a set of measurements can be considered as the sum (or average) of a series of random and independent events. Consider tossing 20 coins, for example. Assuming the coins are fair and unbiased, then whether each coin lands heads or tails is a random event unaffected by other coins. If we wanted to predict how many of the coins will land heads up then a sensible guess is 10, because there are 20 coins and the probability

---

[9]If you are interested in the formula, you can find it explained in Harris and Jarvis (2011: Chapter 3) or by typing 'Normal distribution' into Wikipedia.

of landing heads is $P(H) = 0.5$. However, 10 is an expectation, not a guarantee. In fact, there are 184,756 different ways that 10 out of 20 coins can land heads up. But there are also 167,960 ways for there to be nine heads, 125,970 for eight heads, 77,520 for seven heads, ... and one way for there to be no heads (if all the coins land tails up).[10]

Figure 5.7 displays this information for all the possible outcomes. Note that the values have a symmetrical distribution around the expected value: the number of ways of obtaining 9 heads is the same as the number of obtaining 11 heads, 8 heads is the same as 12 heads, and so forth. Note also that the distribution looks very much like the normal curve. The more coins we toss, the more it will do so.

Now, instead of coin tosses, imagine that the height of a person is the outcome of a lot of somewhat unpredictable factors: the sexual attraction of parents, their height, diet, exposure to disease, environmental conditions and so forth. The net effect of these essentially random events is that people's heights vary in such a way that if you take a reasonably sized sample of other students in your class, plot their heights and look at the distribution, you are likely to find it is normal (or approximately so, at least). There is no great magic to this. When you obtain a normal distribution what it is showing is that values close to the expected value, which is the mean, are more prevalent in the data than values that are unusually high or low. Put another way, the further we step away from the mean, the less probable the values become; or, even simpler, the more extreme the value, the rarer it is.

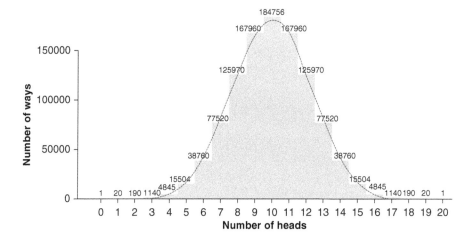

FIGURE 5.7 Showing the number of ways $n = 20$ coins can land with k of them heads side up

---

[10]The equation to calculate this is in Section 4.9.

5.6 SKEWED DATA

Despite its title, the normal distribution is surprisingly elusive when it comes to real-world data sets. Consider the distance to school data: both Figures 5.4 and 5.5 show the data are not symmetric but have a long tail of higher distances that are less common but present in the data nevertheless. This is an example of positive skew, when the tail is to the right of the mean. An indication of positive skew is when the mean is greater than the median for the data because the mean is pulled up by the presence of high values.

Positive skew occurs quite often. US earnings data have a positive skew because of the relatively few very high earners and because of the relatively many lower earners. So, too, do unemployment rates. It happens when the range of possible measurements is bounded only at the minimum: it is impossible for a student to travel less than zero distance to school, but there is not, in principle, an upper limit. Earnings also cannot be negative but, judging from the incomes of the super-rich, the highest incomes keep increasing. Unemployment rates are a little different in that they are bounded on both sides – they cannot be less than 0 per cent nor greater than 100. However, in well-functioning economies the mean is not in the middle of that range. In April 2015 the unemployment rate in the United States was 5.4 per cent. Consequently, the bound below the mean is a lot closer than the bound above it, allowing the above average values to be more spread out.

The opposite of positive skew is negative skew, when the mean is less than the median. Both are shown in Figure 5.8. Sometimes it is possible to increase the normality of the data set by applying a mathematical transformation to it. Often the square root or logarithm of the values is used for positively skewed data. More generally, we might try using $f(x) = x^k$, where $f(x)$ represents the transformed

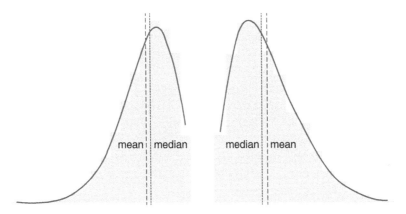

FIGURE 5.8   Examples of negatively skewed data (left, where the mean is less than the median) and positively skewed data (right, the mean is greater than the median)

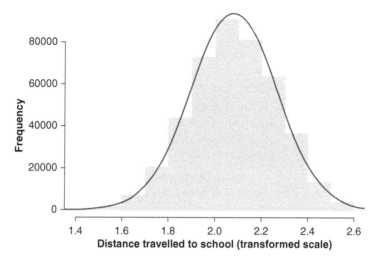

FIGURE 5.9 Applying a mathematical transformation has normalised the distance to school data

data, with $0 < k < 1$ for positively skewed data (e.g. the square root, $k = 0.5$) and $k > 1$ for negatively skewed data. However, this procedure will work only if all the values are positive, so it may be necessary to make them so by raising them by a constant amount: $f(x) = (x + c)^k$. Figure 5.9 shows the same data as in Figure 5.4 but transformed with the constant of $c = 1$ and the power of $k = 0.098$. The positive skew has gone.[11]

5.7 NORMALITY AND STATISTICAL TESTING

The assumption of normality appears a lot in statistical testing. However, it is not necessarily the data that are assumed to be normal but the sampling distribution of a given statistic. To understand this, assume that instead of the (near) complete set of pupil data for London we instead had a random sample of 1000 of those pupils. The mean distance to school for the sample is 2.50 km. It should be obvious that this result is dependent upon whom we happened to sample, and that a second random sample is unlikely to produce the exact same result. In fact, it gives a mean of 2.47 km, while a third gives 2.42 km. We are seeing variation in the sampling statistic, which is the sample mean. Repeating for a thousand samples, further variation emerges, with the range of means extending from a minimum of 2.16 km to a maximum of 2.74 km. There is nothing special about this range – it

---

[11]If you are wondering how I knew the value should be $k = 0.098$, then the answer is that I didn't. Instead I used an optimisation procedure on my computer to work through a range of possible answers and determine which one was best.

will change with the samples. What is more interesting is to plot the sample means and see how they are distributed, which is what Figure 5.10 allows. It shows the sampling distribution of the mean. Note that it is almost normally distributed, despite the fact that the data sampled were themselves positively skewed. This is an example of the central limit theorem, which states that the sum and average of a large number of independent and identically distributed random variables (the sample means, in this example) will generate an approximately normal distribution regardless of the underlying distribution. Moreover, the mean value at the centre of Figure 5.10 is very close to the true average distance travelled to school by pupils in London (that is, close to the value for all pupils and not just those sampled). It is 2.48 km, as is the true value. More precisely, it is 2.482 km whereas the true value is 2.479 km. This is an example of the law of large numbers, which states that as the number of independent and identically distributed random variables increases, their mean gets closer to the true or expected value.

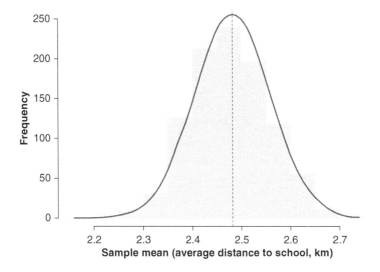

FIGURE 5.10 The central limit theorem and the law of large numbers ensure that, with a sufficiently large sample, the sampling distribution of the mean is approximately normal and centred very close to the true or expected value

5.8 CONFIDENCE INTERVALS

Whilst it is all very well and good to show what happens with 1000 samples, what should we do when there is only one sample of data and that is all the information we have about the population from which the data have been sampled? Rarely are we in a position to keep sampling and show that the distribution of the sample means is centred on the 'true mean' for the population. After all, the very reason for sampling is usually because the complete population is impossible, too expensive or too destructive to measure in its entirety.

Despite this, even with one random sample we know seven things. First, we know its sample mean. Second, we know that mean must lie somewhere on the sampling distribution that would arise if we could sample over and over again. Third, because of the central limit theorem (and because the sample size is not too small), we know that the sampling distribution would be normal. Fourth, because of the properties of the normal curve, we know that 95 per cent of the sampling distribution will be within 1.96 standard deviations of its centre, and 99 per cent within 2.58 standard deviations. Fifth, because of the law of large numbers we know that the distribution is centred on the unknown mean average for the entire population. Sixth, it must follow that 95 per cent of all possible sample means are within 1.96 standard deviations of that population mean, and 99 per cent within 2.58 standard deviations. Therefore, seventh, there is a 95 in 100 chance that the mean for the sample we have is within 1.96 standard deviations of the unknown population mean, and a 99 in 100 chance it is within 2.58 standard deviations. We can know that without knowing the actual value of the population mean.

This is brain aching stuff, and the logic of it is based on us knowing the standard deviation of a hypothetical sampling distribution that would arise if we took very many samples, calculated their means, and looked at how those means were distributed. We don't know that but it can be estimated. Before explaining how, it is helpful to refer to the standard deviation of the sampling distribution by another name, which is the standard error. This distinguishes it from a more usual understanding of the standard deviation as something that measures how spread-out any one set (or sample) of data are around their mean. The distinction is useful because, in fact, one is estimated from the other:

$$se_{\bar{x}} \cong \frac{s}{\sqrt{n}} \tag{5.9}$$

where $se_{\bar{x}}$ is the standard error of the mean, s is the standard deviation of the sample and n is the size of the sample (the number of observations it contains). Given this, and denoting the unknown population mean as μ, we are now in a position to say that

$$P\left(\bar{x} - 1.96se_{\bar{x}} < \mu < \bar{x} + 1.96se_{\bar{x}}\right) = 0.95,$$

$$P\left(\bar{x} - 2.58se_{\bar{x}} < \mu < \bar{x} + 2.58se_{\bar{x}}\right) = 0.99 \tag{5.10}$$

All this amounts to is an algebraic and probabilistic re-expression of the seventh point, above: of all the sample means that might have been obtained, there is a probability of 0.95 that the one we have is within 1.96 standard errors of the unknown population mean, and a probability of 0.99 that it is within 2.58. Bringing this all together, from a random sample of data, and from knowledge about its size, mean and standard deviation, the standard error can be calculated. That in turn allows an interval to be determined that is likely but not certain to contain the population mean. The range $\bar{x} \pm 1.96se_{\bar{x}}$ is the 95 per cent confidence interval. The range $\bar{x} \pm 2.58se_{\bar{x}}$ is the 99 per cent confidence interval.

Looking at how the standard error and the confidence intervals are calculated, they can be understood as measuring the uncertainty of the data. The standard error decreases and the confidence intervals narrow, the larger the size of the sample. That makes sense: a larger set of data is less likely to be affected greatly by rogue values than a smaller one (because the larger set contains more observations, the impact of unusual values is diluted). The standard error increases and the confidence intervals widen, the greater the standard deviation of the data. That also makes sense: more variable data are noisier, so harder to pin down. Finally, the 99 per cent confidence interval is wider than the 95 per cent interval, for the good reason that the wider it is the more confident we can be that it will contain the population mean.

For the first sample of the London pupils data (discussed in Section 5.6), the mean is 2.497 km, the standard deviation is 2.249 km and the number of observations is 1000. The standard error is therefore 0.071 ($=2.249 / \sqrt{1000}$) and the 95 per cent confidence interval is from 2.426 ($= 2.497 - 1.96 \times 0.071$) to 2.568 km ($= 2.497 + 1.96 \times 0.071$), which may be written as 95% CI [2.426, 2.568] and says that we are "95% confident" that the true average distance travelled by London pupils to school is between 2.426 and 2.568 km.[12] But how about the complete set of London data: does it make sense to calculate a confidence interval for that? Certainly it can be done. The 95 per cent confidence interval suggests a mean distance to school of 95% CI [2.473, 2.486] kilometres. The interval is narrow because the data set is so large. The deeper question is what the confidence interval now indicates. It is true that the London data are not actually the complete population: there are values missing – those pupils whose place of residence is not known or those who attend fee-charging schools, for example. However, it is stretching credibility to claim those values are missing at random and that the data can therefore be treated as a random sample in the way the use of confidence intervals traditionally assumes. Even so, there may still be value in providing the confidence interval. This is because it can be argued that there are two roles for confidence intervals in applied data analysis. The first relates to random sampling from a population and an indication of the range within which the unknown population value probably falls. The second is to provide appreciation of a derived value given the noisiness of the

---

[12] This is what the American Psychological Association style guide recommends.

data, sending an important signal about uncertainty and the presence of stochastic errors even in data that purport to be a complete census of a population, for example (Johnston et al., 2014a). The second role remains even where the first does not. This is not, however, a position that all commentators would find credible.[13]

5.9 THE *t* DISTRIBUTION

The validity of the central limit theorem leading to a normal sampling distribution relies on the sample being of sufficient size. The threshold is usually taken to be 30 or more observations, but this is a rule of thumb. For smaller samples, drawn from an approximately normal population, a slightly different distribution needs to be considered, which is the *t* distribution. The shape of the *t* distribution is very similar to that of the normal distribution but has more of its area further from its centre. For a (stupidly) small sample of only five observations, the 95 per cent confidence interval increases from 1.96 standard errors either side of the mean under a normal distribution, to 2.78 under the a *t* distribution. The *t* distribution is wider. How much wider depends on the degrees of freedom, equal to one less than the number of observations. For a sample size of 20 observations and 19 degrees of freedom, the confidence interval is 2.09 standard errors either side of the mean. For 30 observations and 29 degrees of freedom, it is 2.05. The larger the sample size, the more similar the normal and *t*-distributions become (see Figure 5.11). The distribution is used for *t*-tests, statistical tests of difference (see Chapter 6).

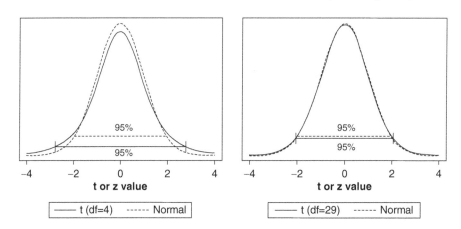

FIGURE 5.11 The *t* distribution is 'fatter' than the normal distribution and leads to wider confidence intervals, although as the sample size and degrees of freedom increase, the two distributions become increasingly similar

[13]The author to whom Johnston et al. (2014a) are responding, for example: Gorard (2014).

5.10 CONCLUSION

The chapter has developed may of the statistical principles found in Chapter 3, focusing on ways to describe and to compare data sets, and introducing concepts such as the degrees of freedom, robust statistics, the normal distribution and z-values. In the later sections, it moved towards statistical inference – the process of using descriptive statistics about the sample to obtain estimates about the population from which the sample was drawn. In particular, the idea of confidence intervals was introduced.

Although this chapter has focused on two of the more common distributions – the normal and t distributions – there are plenty of others that crop up in statistical work. These include the chi-square distribution used for the chi-square test (see Chapter 6), Fisher's F used for analysis of variance (whether the differences between groups are statistically significant with respect to the differences within groups,[14] and in comparing statistical models: see Chapters 9 and 10), the logistic or logit distribution used for modelling a variable with a binary outcome (see Chapter 10), and the Poisson distribution used to model the counts of rare but independent events and also in the description of random spatial point patterns (see Chapter 11).

Statistical inference is linked to statistical testing. For example, using the knowledge we have of two samples to make a judgement about whether they were drawn from the same population. Statistical testing is the focus of the next chapter. It is long established but, as we shall see, controversial.

---

[14]See, for example, Harris and Jarvis (2011: Chapter 6) or an online learning resource such as www.statstutor.ac.uk.

6

STATISTICAL TESTING, STATISTICAL SIGNIFICANCE AND WHY THEY ARE CONTENTIOUS

6.1. INTRODUCTION

Methods of statistical testing are found in almost any textbook introduction to statistics. However, traditional ways of conducting and evaluating statistical tests are increasingly questioned, especially in the social sciences, when the data do not constitute a random sample of a population. Blithe conformity to orthodox practice has been denounced as bad science that can misdirect policy, waste money and, in some circumstances, cost lives. A particular proposition has proved particularly problematic. Called the p-value, it has provoked pages of polemic in print.

The basic idea of statistical testing is simple enough. It is used to test whether some level of difference (or association) between two or more variables has arisen by chance. It may be, for example, that variable X_A has a mean average of 12.4 whereas variable X_B has a mean of 13.6. Clearly they are not equal, but the numeric difference does not, in itself, provide evidence that the variables are measuring categorically different things. Stochastic (random) errors mean that samples of the same population will yield different measurements even though they are measuring the same thing. Judgement on whether they are truly different is based on how far apart the values are (the observed difference), the amount of variation in the data (how noisy they are), and the number of observations, which is the amount of evidence available to make a decision.

This seems logical, so why has statistical testing proved contentious? To understand the debate, this chapter illustrates the thinking with reference to two tests that appear a lot in geography and in other subjects too: the chi-square and *t*-tests. The process of statistical significance is presented before discussing why it is now regarded as flawed. The language of statistical testing is somewhat arcane, which does not aid understanding. However, important concepts are raised that appear again in later chapters and which help explain what has been called the new statistics, briefly discussed towards the end of this chapter.

6.2 THE CULT OF CHI?

Chi-square is a statistical test described by the statistician, Michael Crawley (2007: 222), as 'introduced to generations of school children in their geography lessons, and comprehensively misunderstood thereafter'. It's a mischievous jibe, but not without foundation. In some parts of high school geography, chi-square has almost protected status, remaining present in a number of geography curricula over the decades. Here it is used to demonstrate the process of hypothesis testing, which is one of looking at the data, calculating a test statistic, considering the rarity of the test statistic under an assumption of random sampling, and using that information to determine whether there is something 'significant' between the variables. Chi-square is also used in some spatial statistics (see Chapter 11).

6.3 ABOUT CHI-SQUARE

The thinking behind the chi-square test can be demonstrated with the results of a 2012 survey of geography undergraduates within my own university. Amongst other options, the students were asked whether they identified more as scientists or as social scientists, and also whether they agreed or disagreed with the statement that 'learning quantitative methods is more important for scientists than for social scientists'. To keep things simple (and with apologies to those geographers who identify more with arts and humanities), assume there are only two disciplinary identities, scientist and social scientist, and that the students could only agree or disagree with the statement.

What we have are two variables, each of which consists of two groups or categories. They are categorical variables. There are two types of geographer, scientist or social scientist, either of whom can agree or disagree with the statement. This gives four possible responses: scientists who agree, scientists who disagree, social scientists who agree, and social scientists who disagree. The two variables can be laid out in what is known as a contingency table, where the rows represent the categories for one variable and the columns represent the categories for the other. The number of students in each category can then be counted and displayed in the table.

The purpose of the chi-square test is to consider whether the numbers in the contingency table provide evidence that one variable may be used as a substitute for the other. To appreciate how that could happen, imagine that students identifying as scientists always take the opposite view from those identifying as social scientists and vice versa. This would be the case if, for example, all the scientists agreed with the statement about learning quantitative methods and if all the social scientists disagreed. In such a circumstance, the two variables are completely exchangeable: we know from their disciplinary identity whether a student supports the statement or not, and we know from whether they support the statement whether they identify as a scientist or social scientist. In this situation, the table of counts could look like Table 6.1.

TABLE 6.1 In this hypothetical example, one categorical variable (disciplinary identity) fully predicts agreement or disagreement with the statement 'learning quantitative methods is more important for scientists than for social scientists'

	Agreed	**Disagreed**	**Σ**
Scientists	33 (100%)	0 (0%)	33 (100%)
Social Scientists	0 (0%)	35 (100%)	35 (100%)
Σ	33	35	68

The opposite case occurs when the scientists are no more likely to agree (or disagree) with the statement than the social scientists are. An example of this is shown in Table 6.2. Here the two variables are unrelated and therefore independent of one another. They are unrelated because the percentages of scientists who agree and disagree with the statement are equal to the percentages for social scientists. Consequently, it is not possible to predict a student's attitude towards quantitative methods from their disciplinary identity, nor their disciplinary identity from their attitude.

Tables 6.1 and 6.2 illustrate two extremes. The first is of the variables being fully exchangeable. The second is of them being fully independent. The more

TABLE 6.2 In this example, the two variables are independent of one another because the percentage of scientists who agree/disagree with the statement is the same as for social scientists

	Agreed	**Disagreed**	**Σ**
Scientists	24 (75%)	8 (25%)	32 (100%)
Social Scientists	27 (75%)	9 (25%)	36 (100%)
Σ	51	17	68

usual circumstance is somewhere in-between. The chi-square test sets out the data in a contingency table, compares the table with what it would look like if the two variables were independent, and offers a statistical assessment based on that comparison.

The observed values

So far the examples have been hypothetical. Let's take a look at the actual data from the survey of geography students. The number in each category is shown in Table 6.3. From it we can see that a greater number of social scientists disagreed with the statement than did scientists. That difference could be because the social scientists are more defensive against the idea that quantitative skills matter only to science. I would like to think so but this is not a balanced sample. There are slightly more social scientists than scientists amongst the respondents, so more to disagree. To proceed, what we need to do is compare the observed counts with what they would look like given an expectation of independence.

TABLE 6.3 The actually observed values: the numbers of students who identified as scientists or social scientists and whether they agreed with the statement about learning quantitative methods

	Agreed	Disagreed	Σ
Scientists	15 (45%)	18 (55%)	33 (100%)
Social Scientists	6 (17%)	29 (83%)	35 (100%)
Σ	21	47	68

The expected (or comparative) values

The chi-square test asks whether there are patterns in the data – for example, scientists agreeing, social scientists disagreeing – or whether the variables are independent of each other. Table 6.2 showed that independence is achieved when the percentage agreement/disagreement is the same for the social scientists as it is for the scientists. When this happens, the students' disciplinary identity is indeterminable from their attitude towards quantitative methods. Most statistical software will calculate the expected values for you. To do so manually, consider that there are 68 respondents, of whom 21 agree with the statement about learning quantitative methods. If those agreements were shared out in proportion to the number of scientists and social scientists in the survey, then the number of scientists who agreed would be a proportion equal to 21/68 of the 33 scientists (i.e. $21/68 \times 33 = 10.2$), whereas the number of social scientists would be 21/68 of the 35 social scientists (= 10.8). When the responses are shared out in this way, the percentage agreement for the two groups is the same, at 31%.

Applying the same logic to the 47 students who disagreed, 22.8 would be scientists (= 47/68 × 33) and 24.2 social scientists (= 47/68 × 35). The percentage disagreement is now the same for both groups, at 69%. All these expected values are shown in Table 6.4. Note that the column, row and overall totals are the same as for Table 6.3.[1]

TABLE 6.4   The expected values if the variables were independent of one another

	Agreed	Disagreed	Σ
Scientists	10.2 (= 21/68 × 33) (31%)	22.8 (= 47/68 × 33) (69%)	33 (100%)
Social Scientists	10.8 (= 21/68 × 35) (31%)	24.2 (= 47/68 × 35) (69%)	35 (100%)
Σ	21 (31%)	47 (69%)	68

The test statistic

Now that we have the observed and expected values, a simpler definition of chi-square is that it tests for whether the observed values differ by a statistically significant amount from the expected ones. That difference is formularised by an equation, which is used to generate the test statistic. In our specific example, each of the shaded cells in Table 6.3 is compared with its counterpart in Table 6.4: top left with top left, top right with top right, and so forth, giving four pairwise comparisons. Each expected value (E) is subtracted from its corresponding observed value (O), with the difference then squared (removing negative values) and divided by the expected value. The results are summed together. This calculation is expressed more eloquently in notation than in words, where the chi-square statistic, x^2, is

$$x^2 = \sum \frac{(O-E)^2}{E}$$ (6.1)

which, for our data, gives

$$x^2 = \frac{(15-10.2)^2}{10.2} + \frac{(18-22.8)^2}{22.8} + \frac{(6-10.8)^2}{10.8} + \frac{(29-24.2)^2}{24.2}$$

$$= 6.35$$ (6.2)

---

[1]They have to be, it is from these totals that the expected values are derived.

The result is the test statistic, here equal to 6.35.[2] The greater the test statistic, the greater the observed values differ from expectation. That expectation was formed under a 'what if?' scenario of independence for the variables. If we take that as the default position to stick to until the evidence stacks up against it, then the test statistic provides the gauge for whether to maintain the assumption of independence or reject it. In statistical language, the default is called the null hypothesis (null because it assumes there is no association between the two variables). The alternative hypothesis is its opposite (that there is an association).

In most legal systems, the null hypothesis is an assumption of innocence; the alternative hypothesis is one of guilt. Somebody is declared guilty only when it can be proved beyond all reasonable doubt. This places the burden of proof on the prosecution. The idea is to avoid the null hypothesis being rejected when it is, in fact, true — to avoid sentencing someone for a crime they did not commit. Hence, jurors are required to have a high degree of confidence before reaching a guilty verdict. In a similar way, and to complete the statistical process, we need to determine whether the test statistic provides sufficient evidence to reject the null hypothesis, and thereby judge the variables to be related. The classic way to make this decision is to look at the magnitude of the test statistic and ask whether the result can be attributed to chance (a consequence of random sampling errors). With reference to the chi-square distribution, the probability of obtaining a test statistic of 6.35 or greater, 'by chance' when the null hypothesis is correct, is $p = 0.012$ (1.2 per cent). That is low and leads to the conclusion that the scientists and the social scientists differ statistically significantly in their response to the question 'learning quantitative methods is more important for scientists than for social scientists'. Admittedly, it's not an Earth-shattering conclusion, but the same test could be used to examine whether attitudes to maternity pay vary by gender, whether attitudes to education vary by ethnicity, to measure differences in the outcomes of environmental processes or to test a geographical pattern of events against a hypothesis of complete spatial randomness (see Chapter 11).

6.4 ABOUT THE P-VALUE

The reason for discussing the chi-square test is that it provides an introduction to a number of statistical ideas, including contingency tables, whether two variables are related, the comparison of what we have with some hypothetical 'what if?' and the formalisation of that comparison into a statistical test. Of particular interest — and most contentious — is the p-value. Recall that it provided the motivation for rejecting the null hypothesis and declaring the result statistically significant. In statistical

---

[2]There is another version of this formula that makes a slight correction (Yates's continuity correction) for a 2 by 2 table. However, it is generally regarded as an overcorrection.

testing it is common practice to say a result is significant if the *p*-value drops below 0.05, 0.01 or even 0.001. The lower the threshold, the more stringent the criterion for what qualifies as statistical significance. Each of these is ultimately an arbitrary cut-off, but they imply that the evidence is against the result being due to chance, at a ratio of 95 to 5 against at the first threshold, 99 to 1 against at the second, and 99.9 to 0.1 at the third. That is why they are said to provide 95 per cent, 99 per cent or 99.9 per cent confidence in the result.

Before discussing the limitations of this thinking, it is helpful to understand how the value of $p = 0.012$ was obtained. In short, my computer produced it, as will yours: when you run a statistical test, the software will usually generate a *p*-value alongside the test result. A longer answer is that for any test there is a range of values that are possible but not all equally likely, in the sense that some values are exceeded more often than others. Figure 6.1 shows the test score of 6.35 and the probability it will be exceeded by a larger chi-square value. It rarely is. The probability it will be exceeded is the *p*-value, 0.012. Because it is less than 0.05, the result is considered statistically significant at a 95 per cent confidence level (but not at either 99 or 99.9 per cent). Put another way, the result is fairly rare and unlikely to be a fluke. If the null hypothesis is correct and the two variables are independent of each other, then it is bad luck to get a test statistic that said otherwise because of who happened to be sampled.

The *p*-value is directly related to the test statistic. As Figure 6.1 shows, the greater the chi-square statistic, the lower the *p*-value. However, a second piece of information is required to calculate the *p*-value because it is an oversimplification to talk about 'the' chi-square distribution. Like the *t*- and normal distributions seen

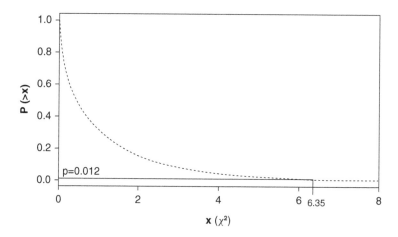

FIGURE 6.1   The probability of exceeding our test score of 6.35 is, in principle, $p = 0.012$. Since this is less than $p = 0.05$, the result conventionally is regarded as statistically significant, at a 95 per cent confidence level

in Chapter 5, chi-square is actually a family of distributions. The *p*-values shown in Figure 6.1 are for the member of that family with one degree of freedom.

Chapter 5 introduced degrees of freedom as the amount of flexibility to change a set of values given that some things are fixed. To give an example, look back at the contingency table in Table 6.3 and assume that the number of agreeable scientists rises by 2. If the row and column totals are fixed, then knowing the number of scientists who agreed means we can calculate all the other values too, as Table 6.5 shows. What this means is that for a 2 by 2 contingency table, we can only change one thing before the rest must follow: the test has one degree of freedom.

TABLE 6.5   As soon as one value in the table is changed, all the other values must 'fall into line' if the column and row totals are fixed

	Agreed	Disagreed	Σ
Scientists	17 (increased by 2)	16 (= 33 − 17)	33
Social Scientists	4 (= 21 − 17)	31 (= 35 − 4)	35
Σ	21	47	68

It is not a requirement for the chi-square test to have one degree of freedom. Variables can be used that consist of more than two categories. In such cases, the degrees of freedom are calculated as (number of columns − 1) × (number of rows − 1).

When the degrees of freedom change, so too does the *p*-value. A part of Figure 6.2 shows what we know already: a test statistic of 6.35 is statistically significant at the 95 per cent threshold ($p = 0.012$) when there is one degree of freedom. It remains significant when there are two degrees of freedom ($p = 0.042$; for a 3 by 2 contingency table, for example). However, it is insignificant ($p = 0.096$) when there are three degrees of freedom, or more. This fits with intuition: having a greater number of categories over which to distribute the numbers provides greater opportunity for the observed and expected values to differ; it is therefore necessary to have more difference and a greater test statistic before it is considered statistically significant.

6.5 PROBLEMS WITH THE *P*-VALUE

Put to one side the equations, the graphs, the tables, the Greek symbol and most of the statistical nomenclature. None of these is especially necessary to understand the rudiments of what has been done, which is to compare two variables using a statistical test and to take the associated *p*-value as the basis for deciding whether the test result is statistically significant. This is standard practice in much social and

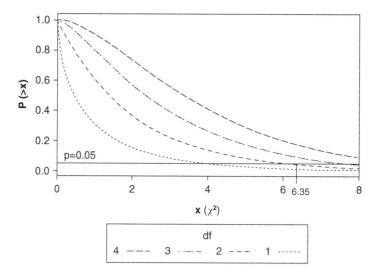

FIGURE 6.2 The cut-off for statistical significance is a function not only of the test score but also of the degrees of freedom (df) for the test. With one or two degrees of freedom, a test score of 6.38 is significant at a 95 per cent confidence (because $p < 0.05$) but not with three or more degrees of freedom

scientific research. Different tests are used, depending upon the data and what exactly is being tested, but the underlying logic of using the p-value to judge whether the result is statistically significant or not remains the same. The problem is that the logic is not as robust as its regular use implies.

What are we sampling?

Recall the words 'for any test there is a range of values that are possible but not all equally likely, in the sense that some values are exceeded more often than others'. The statement suggests that if we reran the survey with a different group of students, we would get a different test result. That, in turn, implies that of all the students who could have answered the survey only some have so far done so, or – to put it another way – that the current respondents are a sample of some larger population of students. So far, so good. The problem lies in the assumption that those who took the survey are a random sample of that larger population. It is the idea of randomness that gives rise to the idea of a sampling distribution and to the idea that the test statistic could, in principle, be exceeded by chance, if it were possible to take a second sample, of the same size, from the same population. But who or what is the population we are sampling from? It could be all students taking geography at Bristol in the year of the survey. However, the sample is by no means random. In fact, it was self-selecting. It was up to the students whether to complete the survey or not. Do those who completed the survey hold the same views as those who didn't? We have no way of knowing, but it is a leap of blind faith to treat the sample as though it were random.

It is not unusual to see statistical tests applied to secondary data – that is, data not collected specifically for the analysis but with a more general usage. It also is not unusual to see statistical tests used on data containing missing values where those values are not missing at random but leave the data biased towards particular subgroups of the population. For example, and as noted in Chapter 5, educational data in the UK tend to omit pupils in fee-charging schools, which means that if you wish to study social segregation between schools, the data are missing about 7 per cent of the more affluent pupils. In other cases, such as with the survey of students but also in the analysis of social media or other sources of 'big' or volunteered data, it is a self-selecting sample. That does not make it uninteresting or unworthy of analysis, but to treat the data as a random sample of a larger population is somewhat incongruous. At the very least, there is need to think carefully about who or what that population is.

Statistical significance and sample size

A second problem with the p-value is demonstrated by multiplying each of the values in Table 6.3 by 10, which is how they appear in Table 6.6. This raises each of the expected values by 10 and means that, if you do the calculations, you will find the test result is now also 10 times as large, equalling 63.5. The p-value for a chi-square statistic of 63.5 with one degree of freedom is so small that the result appears significant at more than 99.99999 confidence.

What has changed? Not the relationship between the two variables. They are no more or less related than they were previously (compare the percentages in Table 6.6 with those in Table 6.3). The reason the p-value has decreased is that a larger sample brings greater confidence in saying something is or isn't statistically significant. That makes sense: we are basing the decision on more evidence. However, it also confirms what others have noted: 'a significance test is little more than a roundabout measure of how large a sample is' (Murphy and Myors, 2003: 16).

The link between sample size and the statistical significance of a test result is well known, although usually considered in terms of the sample size being too small to detect what is being tested for – a test that is said to lack statistical power. Imagine a retailer wants to suggest there is no difference in consumer preference for their own brand label and some more expensive 'big brand' alternative. One way to achieve this is to ask very few people their preference – too few for any statistically significant differences to emerge.[3] Failure to reject the null hypothesis

---

[3] Of course, that is tantamount to cheating, and any retailer that did set up the test to fail could find itself in breach of advertising regulations were it to publicise the result. Another way to be conservative with the truth is to repeat the test with different groups of people and to keep doing so until you get the result you want.

is not proof that the null hypothesis is correct. It could be a failure to collect sufficient evidence against it.

Various online power calculators are available that can be used to guide the amount of data that is needed for the test to be able to detect something, should it exist. However, in the age of large data sets, a thornier problem is that with lots of data, even the trivial will appear significant. Collect enough data and something 'significant' will emerge. This is not a problem of the chi-square test but with the logic of significance testing.

TABLE 6.6   The original values multiplied by 10. These lead to a chi-square statistic of 63.5 (1 df; $p < 0.001$)

	Agreed	Disagreed	
Scientists	150 (45%)	180 (55%)	330 (100%)
Social Scientists	60 (17%)	290 (83%)	350 (100%)
	210	470	680

6.6 THE *t*-TEST

Another commonly used statistical test, the *t*-test, illustrates the problem of large data sets giving rise to results that, although statistically significant, are also substantively trivial. There are a number of variants of this test, but a frequent use is to compare the mean averages of two sets of data, testing against the possibility that they are both measuring the same thing.

The idea employs many of the principles described in this chapter and also in Chapter 3. Finding that the average for one data set is not equal to the average for the other is not proof that they are from different sources. It is improbable that their means will be exactly the same. Intuitively, it is the amount of difference that matters. The further one average is from the other, the less likely they are both samples of the same population. That intuition is correct, except that two additional pieces of information need to be considered. First, the noisiness of the data, since a highly variable population can create greater variation between samples and their averages. Second, the amount of data, because the more data we have, the more confident we become in detecting a difference. To return to the court metaphor: we have more evidence by which to reach a verdict.

The default assumption for the *t*-test is that the two data sets are measures of the same population. The null hypothesis says that if whatever the first data set is measuring could be measured in its entirety then that complete population would have the same average value as the complete population for whatever the second data set measures. Put more succinctly, the null hypothesis is that the (unknown) population mean for data set A is equal to the (unknown) population mean for data set B. If true, it implies the populations are the same; that the two data sets are measuring the same thing. The problem is that we cannot measure those populations entirely, although we can still use the data we do have to retain or reject the null hypothesis. We are more likely to assume the data are drawn from different populations the greater the difference between their means, the less the noisiness of the data, and the greater the amount of data upon which to reach a decision.

These considerations all come together in the formula for the *t*-test, which could be written as

$$t = \frac{\text{difference between the means}}{\text{noisiness of the data relative to the amount of data there is}} \qquad (6.3)$$

which, although not a proper expression of the *t*-statistic, has the same elements as a more complete formula such as

$$t = \frac{\bar{x}_A - \bar{x}_B}{\sqrt{s_A^2/n_A + s_B^2/n_B}} \qquad (6.4)$$

where A and B are two data sets, $\bar{x}$ is a mean, s a variance and n is the number of observations in the data set (the sample size).[4] The test has $(n_A - 1) + (n_B - 1)$ degrees of freedom, because for each data set you could hold its mean constant and change all but one of its data values before the remaining value has to fall into line.

Knowing this, imagine we have two data sets, A and B, each consisting of 100 measurements, which, when plotted on a box plot, look like Figure 6.3. Actually, Figure 6.3 is not just a box plot, because it also includes the top half of a violin plot (so called because if the shape of the curve were mirrored on the other side of each box plot, it would look a bit like a violin; see Chapter 7). The point is that Figure 6.3 shows the shape – the distribution – of both A and B, and they look almost the same. In fact, B is just A shifted rightwards a little, so whereas A has a mean average of 100.0, B has a mean of 100.1.

Running a *t*-test comparing these data produces a test statistic of $t = -0.648$, with 198 degrees of freedom. The negative value merely indicates that the mean of A is less than the mean of B. Of relevance to the discussion is the *p*-value of 0.518. From it we conclude that there is no significant difference between the two averages, which seems reasonable given the tiny difference between them. However, consider what happens when A and B keep exactly the same shape and

---

[4]This is not the only way of calculating the *t*-statistic (see, for example, Harris and Jarvis, 2011: 149); however, the principle remains the same.

distance apart, but this time the samples consist of 1000 measurements apiece. The test statistic is now $t = -2.222$ with 1998 degrees of freedom, giving a p-value of 0.026, less than 0.05 and statistically significant at 95 per cent confidence. Raise it to 5000 measurements and a p-value of less than 0.001 is obtained (99.9 per cent confidence). As with the chi-square test, the significance of the t-test value is a function of the sample size.

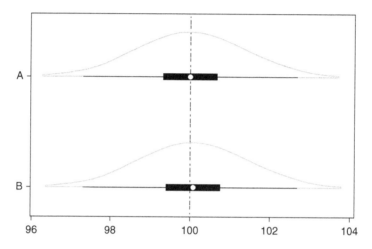

FIGURE 6.3   The only difference between data set A and data set B is that all the values in B are 0.1 greater than their equivalent value in A

6.7 IS STATISTICAL SIGNIFICANCE SIGNIFICANT?

The problem with the chi-square, t and other tests employing statistical significance is that they don't really provide the answer we want from them: they don't tell us whether what we've tested is of any substantive importance or not (whether it actually matters). For the t-test we were looking at a mean difference between the data sets of only 0.1. That might seem trivial, but it depends on the context. Sometimes small differences matter. Sometimes large ones don't.

Given that statistical significance isn't the same as social or scientific importance, that the p-value is a probabilistic abstraction based on the (often) false idea of random samples, and taking into consideration that it's all largely governed by the size of the data set anyway, might it be a good idea to drop the charade? Certainly that is what some authors have argued, some quite vociferously: 'Null-hypothesis significance testing is surely the most bone-headedly misguided procedure ever institutionalized in the rote training of science students' (Rozeboom, 1997, cited by Thompson, 2004). In 2015, a psychology journal, *Basic and Applied Social Psychology*, generated publicity and debate by banning the publication of p-values within its pages.

The argument is not new. The following statement was written in 1978 and was not the first to express concerns: 'I am saying that the whole business is so radically defective as to be scientifically almost pointless' (Meehl, 1978, cited by Thompson, 1996). More recently it has been argued that the cult of significance testing skews knowledge and generates misunderstanding, costing jobs, justice and lives (Ziliak and McCloskey, 2008). Consider that the tradition in academic studies is to publish only statistically significant results. Arguably, this promotes a culture where all people do is look at the p-value, not the meaningfulness of the result. In statistical modelling it is possible – indeed, easy, with large data sets – to have a model full of statistically significant variables that don't actually explain much about what is being modelled. Moreover, if we accept the idea that significant results can occur 'by chance' yet only significant results tend to be commented upon or published, then what arises is the possibility that a freak 'significant' result will grab the headlines whilst the more usual and insignificant ones never see the light of day. Our knowledge is then based on the exception, not the norm.

Consequently, it is hard to defend the practice of statistical testing, perhaps especially in the social sciences where random trials, randomised experimentation and random sampling are rare but not unheard of. Randomised control trials (RCTs) are generally regarded as amongst the best ways to determine cause-and-effect relationships. In RCTs one group of people randomly is selected to receive some sort of intervention (often a medical drug, but it could be an economic incentive to change behaviour, for example) whereas as a comparable and control group is not (or receives a placebo). The idea is that if the selection into groups is unbiased then any difference that is measured between the groups after the intervention ought to be attributable to that intervention alone and not to any systematic difference in the composition of the groups. Unfortunately, aside from the possible ethical or political issues involved (e.g. giving some people a financial inducement and not others), social as well as natural processes are not easily isolated from other confounding factors or so immediately responsive to an input that their causes can be readily determined using an RCT approach. However, RCTs can be very useful for finding ways of nudging people towards particular outcomes. One study showed that homework completion increased by 25 per cent by sending parents personalised text messages when their children did not hand in their work, relative to the completion of children whose parents did not receive such messages.[5]

Even in the 'hard sciences' statistical testing has been discredited: in part because what matters most is the amount of association or difference between two or more variables, not whether it appears statistically significant or not; in part because of

---

[5] http://www.huffingtonpost.com/jon-jachimowicz/randomized-controlled-tri_b_6507928.html

the unhelpful bias towards only publishing statistically significant results; in part because it is somewhat peculiar to consider how often the test statistic would be exceeded if you resampled from the same population *ad infinitum* (but for some reason kept the sample size the same each time); and in part because the whole idea of testing against a null hypothesis presupposes that the null hypothesis was ever plausible or sensible in the first place. Recall that the null hypothesis assumes no difference or association. That is an extreme position to adopt. Arguably the test would better allow for some amount of difference or association between the variables – a null hypothesis of negligible rather than null effect (see, for example, Murphy and Myors, 2003).

6.8 THE NEW STATISTICS (OR, DOING THINGS WITH EFFECT)

Despite the problems with statistical testing, it is still commonplace to see statements of the kind 'the test result is not significant at the 95 per cent confidence level because $p < 0.05$'. This may be true, but it is opaque. At the very least it is sensible to give the actual p-value so that the reader can determine the confidence level it does meet which, as a percentage, is $100 - 100p$.

Another possibility is to state the confidence interval. For the first of the t-tests discussed in Section 6.6, which had sample sizes of 100, the 95% CI for the difference between the means is [−0.404, 0.204]. The fact that this range includes zero indicates that the 'true difference' between the samples could be zero – hence, no significant difference. For the second test, with sample sizes of 1000, the 95% CI is [−0.188, −0.012]. This interval does not include zero – hence, the result is considered statistically significant at 95 per cent confidence. For the third, with sample sizes of 5000, the 95 CI is [−0.139, −0.061]. These confidence intervals are based on the standard error of the difference, not the standard error of the mean that we saw in Chapter 5, and the equation for calculating the standard error is different. However, the principle is the same: the more data we have and the less variable they are, the narrower the interval because the uncertainty decreases.

A problem here is that the choice of the 95 per cent or any other confidence interval is arbitrary. In itself, the 95% CI [−0.404, 0.204] provides no information about whether the difference between the means could be considered statistically significant at a less conservative, 90 per cent confidence, for example.[6] The confidence intervals also remain affected by the sample size, just like the p-value: the more data there are the narrower the confidence intervals will be. The validity or otherwise of applying confidence intervals to samples that are

---

[6] By convention, a 90 per cent confidence interval is less often used, but there is no particular reason why it should be so.

not random has been discussed in Chapter 5. Either way, the root concern remains: the difference between significance in a statistical sense and in a 'the result actually matters' sense.

In the literature there is talk of a new statistics. This advocates a change of emphasis away from null hypothesis tests and p-values to measuring effect sizes. This means looking at the amount of association or difference between data sets rather than relying on statistical tests to judge whether a result matters or not. For the chi-square tests discussed earlier we were interested in whether there is any association – a correlation – between disciplinary identity and support for the statement about quantitative methods. In fact, we don't need any extravagant methods to suggest there is. Looking back at Table 6.3, we can see that the percentage of geographers who identified as scientists and who agreed with the statement 'learning quantitative methods is more important for scientists than for social scientists' is 45 per cent. The corresponding value for the social scientists is 17 per cent. So, for the scientists, agreement is 28 percentage points greater $(= 45 - 17)$ or 2.65 times as great $(= 45/17)$. That seems to be quite a difference. Simply by reporting the results I can leave others to draw their own conclusions, acknowledging the caveat that there is no particular reason why a sample of Bristol students should be representative of other geographers and many good reasons to assume it is not.

A slightly more elaborate measure is the odds ratio, which measures the ratio of agreement to disagreement for one group relative to the other, or – in other contexts – the ratio of success to failure. For the scientists, the observed proportion of agreement is $p_{(\text{Sci, agree})} = 15/33$, which leaves the disagreement at $p_{(\text{Sci, disagree})} = 1 - p_{(\text{Sci, agree})} = 18/33$. For the social scientists, $p_{(\text{Soc Sci, agree})} = 6/35$ and $p_{(\text{Soc Sci, disagree})} = 1 - p_{(\text{Soc Sci, agree})} = 29/35$. The odds ratio, OR, is

$$\text{OR} = \frac{p_{(\text{Sci, agree})} / 1 - p_{(\text{Sci, agree})}}{p_{(\text{Soc Sci, agree})} / 1 - p_{(\text{Soc Sci, agree})}} = \frac{\dfrac{15}{33} / \dfrac{18}{33}}{\dfrac{6}{35} / \dfrac{29}{35}} = \frac{15/18}{6/29} \cong 4.03 \qquad (6.5)$$

The scientists are four times more likely than the social scientists to agree than disagree with the statement.

We can also calculate the correlation, r, between support for the statement and disciplinary identity. One way is

$$r = \sqrt{\frac{\chi^2}{n}} \qquad (6.6)$$

For the data in Table 6.3 this gives $r = \sqrt{6.38/68} = 0.306$ and, for the data in Table 6.6, $r = \sqrt{63.8/680} = 0.306$. Note that the two values are the same. Recall that the data in Table 6.6 are just the original numbers multiplied by 10. The measure of effect size is invariant to sample size, which is true also of the percentage measures and odds ratios but not of the p-value (which isn't a measure of effect size but of so-called statistical significance). A value of $r = 0.10$ can be regarded as a small effect, 0.30 a medium effect and 0.50 a large effect. However, these are just rules of thumb.

There are lots of ways of measuring effect sizes. Measures of correlation are popular for looking at associations (as for the chi-square test and also for correlation and regression, discussed in Chapters 9 and 10). For t-values and looking at the differences between sample means, Cohen's d widely is used. There are plenty of online calculators that can calculate effect sizes. For the data shown in Figure 6.3, where $\bar{x}_A - \bar{x}_B = 0.100$, and with $n_A = n_B = 100$ and $|t| = 0.648$, Cohen's d is 0.092. It is almost the same when $n_A = n_B = 1000$ and $|t| = 2.222$. Although the importance of the effect depends upon the context, effect sizes using Cohen's d are often taken as small (0.2), medium (0.5) and large (0.8). In our case, the 0.1 difference between the sample means may be interpreted as trivial.

6.9 CONCLUSION

This chapter has focused on the chi-square and t-tests to discuss the principles of statistical testing. As previously noted, such testing will appear in almost any introduction to statistics in geography, despite the fact that the enterprise has been thoroughly questioned, not by people who are in some way 'anti numbers' but by those who undertake quantitative research. A casual reliance on the use of p-values has been identified as bad practice in science and in social science, with some calls to ban the linking of the word 'significant' to results that happen to have a low p-value. Statistics need to be interpreted in context, with consideration given to the effect size. Ultimately there are no hard-and-fast rules about what matters. It depends on what is being measured and why. Nevertheless, previous studies may provide guidance, or at least a point of reference and comparison.

We end by noting that the critique is of traditional forms of statistical testing, notably null hypothesis significance testing. It is not a critique of all statistics. The take-home message is to consider what results mean in a real-world sense, not whether they happen to meet some threshold for statistical significance or not. Real understanding is much better than blind obedience to a set of push-button tests and routines.

PART 3
DOING QUANTITATIVE GEOGRAPHY

7

DATA PRESENTATION AND GRAPHICS

7.1 INTRODUCTION

This chapter takes a break from some of the mathematical and statistical intricacies of earlier chapters and gives attention to the visual presentation of data. There are a number of excellent texts on this subject, notably Darrell Huff's classic *How to Lie with Statistics* (1991 [1954]), the various works of the statistician and artist, Edward Tufte, including *The Visual Display of Quantitative Information* (2001) and *Envisioning Information* (1990), William Cleveland's *Visualizing Data* (1993), David McCandless's *Information is Beautiful* (2012) and *Knowledge is Beautiful* (2014), and Stephen Few's *Show Me the Numbers* (2012). These books differ in their approach and focus but have in common a desire to present data effectively in ways that support the cognitive processes of generating knowledge from quantitative information.

According to a blog posting on Naturejobs:

> Data visualisation has become ever more important as the volume of data is increasing. You see data everywhere, in simple infographics, in the sports reporting and in daily news. We are constantly bombarded with it, and are easily confused by it all. Is it because when we go through our studies as young adults we are only presented with a very small variety of charts? This limitation in our understanding of datasets could be helped by using better visuals.[1]

In short, the importance of data presentation has grown with the amount of data collected about people and places, increased access to those data through data stores, data archives and data portals, and new, often free technologies and software that make graphics easy to produce with a few points and clicks or a bit of coding.

---

[1]http://blogs.nature.com/naturejobs/2014/05/06/visualising-data-with-andy-kirk

Despite this, the principles of good presentation are not widely discussed in most textbook introductions to quantitative methods in geography. This is unfortunate, because any student can expect to produce a graph or table at some point in their study yet many will do so badly.

This chapter only scratches the surface of an important and growing area of research and application, but it has a clear message, which is about both analysis and communication: good graphics help you to make sense of data; good graphical design helps others to make sense of them too.

7.2 THE PROBLEMS WITH PIE CHARTS

One of the most widely taught methods of data presentation is also roundly condemned by those with an interest in visual communication. It is the pie chart. The help menu for my statistical software puts it like this: 'Pie charts are a very bad way of displaying information.'[2] You can't get much plainer than that.

Figure 7.1 gives an example. It is supposed to show passenger traffic in the world's five busiest airports in 2013, but is almost entirely incomprehensible. Because it fills almost one quarter of the pie, you might deduce that Atlanta (ATL) has more passengers than the other four, at about 25 per cent of their combined total. However, it's not at all obvious which of London (LHR) or Tokyo (HND) handles the greater number of passengers, nor what percentage of the total those or the other airports take.

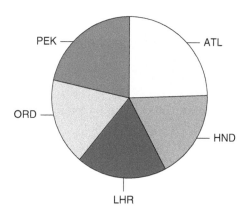

FIGURE 7.1 Pie charts are a bad way of displaying information. In this case, the information is the number of passengers at the world's five busiest airports in 2013 (source of data: http://www.aci.aero/Data-Centre/Annual-Traffic-Data/Passengers/2013-final)

---

[2]The open source software, R.

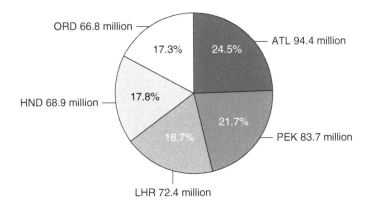

FIGURE 7.2 Adding annotation can make the pie chart more interpretable. In this example, it shows the number of passengers and also that number as a percentage of the total. There are better ways of presenting the data, however

The problems lie with the fact that pie charts work on angles, but angles are hard to interpret and to compare with one another. There are ways to make the chart more intelligible; by sorting the airports clockwise from most to fewest passengers, for example, which makes their rank ordering obvious. Adding annotation makes the values clear too. The result is Figure 7.2. The need to add annotation exemplifies the shortcomings: if the graphic makes little sense without the numbers it is supposed to illustrate, then it can't be a very effective method of data communication.

Doughnut charts

An alternative to the pie chart is the doughnut chart (Figure 7.3). Visually it is more appealing, but since it is only a pie chart with a hole in the middle, all the same criticisms apply. Better alternatives are discussed below.

'3D' pie charts

Before we get to those alternatives, let's consider a true aberration, the so-called 3D pie chart, on the left in Figure 7.4. The impression of depth (the third dimension) is purely a visual effect that has no connection to any data. Because it has no connection, it is superfluous and we lose nothing in taking it away. In fact, its presence only makes a bad chart harder to interpret. There is worse, though: the exploding 3D pie chart, on the right in Figure 7.4. Although a number of well-known software packages tempt you to use these, the best advice is don't, especially not in writing scientific reports or in student dissertations.

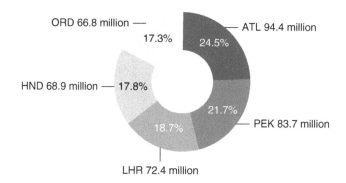

FIGURE 7.3   A doughnut chart is just a pie chart with a hole in the middle so faces all the same criticisms that a pie chart does

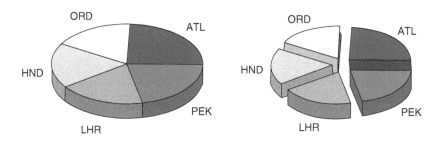

FIGURE 7.4   A 3D pie chart and an exploding 3D pie chart. Just don't!

7.3 BAR PLOTS AND DOT CHARTS

The pie chart has always been an odd choice for displaying data because the eye finds it hard to interpret amounts and differences around a circle. It is much easier simply to line the values up alongside each other, as in a bar plot. Figure 7.5 gives two examples, still using the airport data. I favour the graph on the right. One reason to prefer it is that it has the airports sorted by passenger numbers instead of alphabetically. This makes it easy to see their rank ordering: Atlanta (ATL) is the busiest of the five, Chicago (ORD) the least busy. Second, it labels the vertical axis with the passenger numbers in millions rather than in scientific notation, which will make more sense to most people. It also has those numbers written horizontally, which makes them easier to read. In addition, I've added the numbers for 2012 to compare with 2013.

From 2012 to 2013, passenger numbers increased in all the airports except Atlanta. However, the increase isn't obvious for Chicago (ORD) for two reasons. The first is that the increase was relatively small, from 66.6 million in 2012 to 66.7

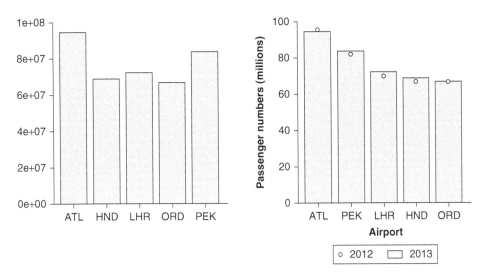

FIGURE 7.5   Bar plots of airport passenger numbers data. The right-hand plot does not rely on the software's default settings and is the better of the two in my opinion

million in 2013. The second relates to the design of the plots. Notice that most of the chart is filled by the bars extending down to zero, although the bit of the graph we are especially interested in is the differences in the values at the top. We might redesign the chart, removing what could be regarded as redundant information, leaving only the upper part instead. The problem with this approach is it makes the chart deceptive. On the left in Figure 7.6 it looks like Chicago has about half as many passengers as Atlanta; on the right it looks like it has less than a fifth: same data, different visual impressions, neither of them correct. In fact, Chicago handles about 70 per cent of the number Atlanta does. That is the message the graph ought to convey, but it will only do so when the bars begin at zero. A popular trick in the news media is to exaggerate differences by choosing a non-zero base.

It is sometimes argued that bar plots are easier to read if they are rotated and drawn horizontally instead of vertically, the argument being that many societies are taught to read from left to right then down and the horizontal bars reflect this. The left-hand part of Figure 7.7 redraws the plot in this way. Leading on from it, the right-hand part of Figure 7.7 takes a more minimal approach, stripping back the bars and leaving only a point symbol to indicate their lengths. It is called a dot chart, to which I have added the arrows to emphasise the increase or decrease in passenger numbers from the one year to the next. Notice that the dot chart does not have an origin at zero, yet, without the bars, does not generate the same visual distortion that the bar plots did when treated in the same manner. To my eyes, the dot chart conveys the information more precisely and clearly. It minimises wasted ink – maximises what Tufte (2001) describes as the data to ink ratio – and is straightforward to interpret.

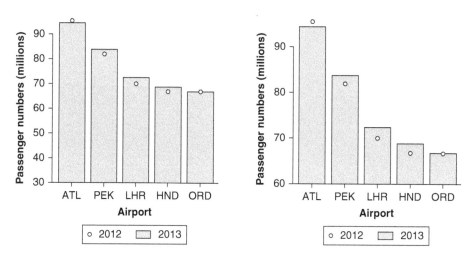

FIGURE 7.6 Starting the vertical axis at values other than zero gives a deceptive impression of the numbers of passengers in each airport relative to each other

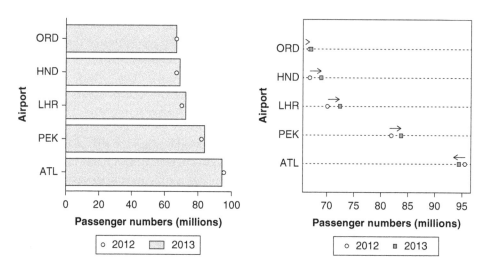

FIGURE 7.7 Bar plots can be easier to read when drawn horizontally, but a more effective way of presenting the same information is the dot chart (right)

Stacked bar plots

A less successful method of presenting the same data is the stacked bar plot, shown in Figure 7.8. The stack is formed by grouping the data into a single block per year (there are two stacks in Figure 7.8 because there are two years). From the plot, we

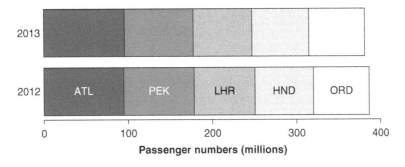

FIGURE 7.8 Stacked bar plots are hard to interpret because it is difficult to work out the length of each part of the stack

can just about see that total passenger numbers were greater in 2013 than in 2012, that Atlanta was the busiest airport and also, if we look very closely, that the number of passengers decreased in that airport from 2012 to 2013. However, it is difficult to read off values for any of the other airports because they don't align with anything in particular but are lumped together as part of a stack. Consider London (LHR), for example. Are its passenger numbers easier to estimate from Figure 7.8 or from either part of Figure 7.7? As with pie charts, software may tempt you into using stacked bar charts, but there are usually better alternatives. Pseudo-3D bar plots are not amongst them.

7.4 TABLES VERSUS GRAPHS

Another alternative to a pie chart, and to a bar chart, is not visual at all, at least not in the sense of being a graphic. It is simply to present the data in a table. Tables can be quickly produced in most word-processing or spreadsheet software, but speed is not necessarily an asset if the end result looks slapdash and rushed. Table 7.1 is unattractive, with its black header column, thick borders and unnecessary stripes.

TABLE 7.1 An ugly table of the airport passenger data

Airport	City	Passengers 2013	Passengers 2012	% change
ATL	Atlanta	94,431,224	95,513,828	−1.133
HND	Tokyo	68,906,509	66,795,178	3.16
LHR	London	72,368,061	70,038,804	3.32566
ORD	Chicago	66,777,161	66,629,600	0.22
PEK	Beijing	83,712,355	81,929,359	2.176

It also lacks consistency: the percentage changes are shown to a varying number of decimal places. Little consideration has been given to the row order and what the reader should learn from the data.

Tables 7.2 and 7.3 are much better. They are the same, except in the way the data are sorted. Table 7.2 emphasises the busiest airport (ATL). Table 7.3 emphasises the one with greatest yearly change (LHR). Neither is wrong or right, but the choice does affect the take-home message of the data. The data are presented tidily and consistently, using three significant figures (three digits, ignoring any leading zeros) in all but one case. The exception is the percentage growth for Chicago. This would appear as 0.221 if the same rule was applied. However, I thought it looked odd for this number to be written with three decimal places (three numbers after the decimal point) when the others in the column appear with only two.

The preference for three significant figures is a matter of personal taste. It doesn't really matter as long as it makes sense in the context of the data and what they are used for. I could write that the percentage change in passenger numbers

TABLE 7.2 A better tabulation of the data, with the rows ordered to emphasise the busiest airport

		Passengers (millions)		
Airport	**City**	**2013**	**2012**	**% Change**
ATL	Atlanta	94.4	95.5	−1.13
PEK	Beijing	83.7	81.9	2.18
LHR	London	72.4	70.0	3.33
HND	Tokyo	68.9	66.8	3.16
ORD	Chicago	66.8	66.6	0.22
	Total	386	381	1.39

TABLE 7.3 In this example the airport with greatest growth is emphasised

		Passengers (millions)		
Airport	**City**	**2013**	**2012**	**% Change**
LHR	London	72.4	70.0	3.33
HND	Tokyo	68.9	66.8	3.16
PEK	Beijing	83.7	81.9	2.18
ORD	Chicago	66.8	66.6	0.22
ATL	Atlanta	94.4	95.5	−1.13
	Total	386	381	1.39

was −1.13345263473 per cent in Atlanta over the period, but the precision is spurious and does not add to the key point, which is that passenger numbers fell slightly over the period. Arguably, rounding degrades the data. For example, the number of passengers for ATL is now recorded at 94.4 million (i.e. 94,400,000) when the true value is 94,431,224. Again, what matters is what we are trying to communicate. Rounding makes it no less obvious that Atlanta is the busiest airport, so it seemed sufficient to write the value to the nearest 50,000.[3] Furthermore, it's much easier to appreciate the scale of the numbers when they are clearly reported in millions.

Because they give the numbers, it could be argued that the tables are more informative than the charts. Certainly tables are more useful if those numbers need to be communicated more precisely than a graphic allows. The disadvantage is that tables are not as visually appealing, can take longer to interpret and can become very long and unwieldy if there is a lot of information to present. Only the most dedicated tablophile will bother to study a table that is many pages long.

7.5 INFOGRAPHICS

A different approach to data presentation has its roots less in scientific visualisation – the methods of illustrating data to help understand and to communicate the results arising from statistical and scientific analysis – and more in data journalism, which is about visualising data to help support and to tell a news story to a general audience (see, for example, Rogers, 2013). An infographic is simply a visual representation of data or information, including a weather or a subway map. Figure 7.9 is a simple example. Were it to appear in a newspaper, I would imagine the circles being replaced with something more interesting, like the shape of an aircraft taking off or landing. Nevertheless, the graphic contains many of the same design elements that appear in the media, where the knowledge to be gleaned from the data is presented by means of annotation, variously sized symbols and their positioning on the chart. The idea is for the graphic to be as self-evident as possible. This approach contrasts with something like a bar plot which, although not difficult to understand, still requires an element of training, which may make it inaccessible or off-putting to a lay audience.

Typing the word 'infographic' into a search engine reveals lots of examples, covering all sorts of topics from air quality to zoology and pretty much everything in-between. Some are very elaborate, dealing with complex data (have a look at the *Guardian*'s visualisations of UK government spending by department);[4] others

---

[3]A rounded value of 94.4 million implies the actual value is somewhere between 94,350,000 and 94,450,000.

[4]The 2011–12 version is at http://www.theguardian.com/news/datablog/2012/dec/04/government-spending-department-2011-12

FIGURE 7.9 A simple infographic of the airport passenger data

seek to get across a political, social or environmental message (view the info-graphics on the True Activist website).[5] A criticism of some infographics is that they are not neutral but advocate on a topic or subject about which the author or organisation campaigns. As such, they are tools of persuasion as much as data presentation. Indeed, many deliberately mix data, presentation and commentary to emphasise a particular cause (Figure 2.4, the graph sceptical of global warming, did that). Whether this matters or not depends upon your point of view. Graphics have long been used to help convey an argument. When John Snow published his famous map of cholera deaths in central London in the second edition of his book *On the Mode of Communication of Cholera* (1855), he was doing so to support a theory developed in the first edition: that cholera is a waterborne, not airborne, disease. The production of the map is not a neutral act by a disinterested and dispassionate scientist, but one with a purpose: to emphasise the deaths that clus-tered around a water pump from a contaminated water supply.[6] In a similar way, an infographic produced for the children's charity Barnado's was designed to emphasise that some of the most vulnerable members of society are young people leaving the care system. It makes no secret of this; it says as much at the top (see Figure 7.10). I can see no problem here: data should be used to illuminate, chal-lenge and cajole. The problem would be if the authors were deliberately dishon-est with the data. Notice that this particular infographic makes liberal use of

---

[5]http://www.trueactivist.com
[6]That map and lots of other interesting maps of London can be viewed at http://mapping london.co.uk/2011/john-snows-cholera-map/

Helping children move beyond care

Young people leaving the care system are some of the most vulnerable members of our society, and Barnardo's is always there for them

Children leaving care have to manage a life on their own much sooner than other young people. Many find themselves living in unsafe, unsuitable or even dangerous B&Bs for long periods of time. Barnardo's is campaigning to make sure young people leaving care get the support they deserve to find somewhere suitable and safe to live.

of 16- and 17-year-olds in the UK do not live with their parents **7%**

of 16- and 17-year-old care leavers in England live alone **33%**

Local authorities often place care leavers looking for a place to live in unsuitable B&Bs

73%
of local authorities in England used B&Bs to place care leavers in 2013-14

46%
of local authorities used B&B accommodation repeatedly

51%
of local authorities in England placed care leavers in B&Bs for 28 days or more, in breach of government guidance

Barnardo's is making the difference

of Barnardo's services for care leavers in England provide support with accommodation needs **60%**

In the 12 months to July 2014 Barnardo's provided support to

2,000

care leavers in England through the charity's leaving care services

Supporting the unsupported

Ever since Dr Barnardo set up the first Ragged School in 1867, Barnardo's has worked to one philosophy – that all children deserve the best start in life. Today, alongside our UK-wide service provision, we also speak out and seek changes in policy and practice that will improve the lives of the most vulnerable children, young people and their families. Our campaigning relies on the involvement of our supporters, and we are always looking to the public to add their voice, and increase the impact that we can have. Find out more at **barnardos. org.uk/beyond-care**

Believe in children
Barnardo's

FIGURE 7.10 An infographic produced by the *Guardian* newspaper in association with the children's charity Barnardo's (source : http://www.theguardian.com/barnardos-support-children/graphic/helping-children-move-beyond-care)

doughnut charts, and half-doughnut charts, also known as gauge charts. This emphasises a difference in approach between academic report writing and the more scientific visualisation on the one hand, and data journalism and graphic design on the other. The first aims to present data and the information it contains as accurately and precisely as possible with minimum distortion, whereas the second aims to present the data in a vivid and accessible way, in order to help tell a story. For examples of the latter approach, have a look at David McCandless' website,[7] or see McCandless (2012, 2014).

Word clouds

A popular but overused infographic of recent years is the word cloud or wordle. It highlights the frequency of words that appear in a document such as the transcript of a political speech, giving a visual impression of its main themes. At its simplest, the production of a word cloud involves counting the number of times each distinct word appears in the document and then plotting those words in a size proportional to the frequency. In practice, the process requires cleaning the data, such as converting upper case letters to lower case, removing stop words (common words like 'a' or 'the') and possibly stemming (reducing words to their stem so that 'mapping', 'maps' and 'mapped' become 'map'. A word cloud is a naïve form of textual analysis since words lose meaning out of context – it becomes unclear whether the word has a positive or negative usage in the original text, for example – and it is often phrases and other combinations of words that are relevant, not single words on their own. Nevertheless, online tools such as www.wordle.net make word clouds easy to produce and distribute.

Figure 7.11 gives four examples. They analyse tweets from or to @BBCNews, @CNN, @FoxNews and @AJEnglish (Al Jazeera) at about 3.15pm GMT on 14 July 2015, the day when a deal was reached with Iran concerning nuclear activity and the lifting of sanctions. I shall not say which is which – see if you can identify the news channel from the word content. Each has been produced slightly differently to show some of the possible methods of design. In the upper left, the allocation of colours to words is random; for the rest, the most frequent words are coloured darkest. The positioning of the words is random for both of the upper charts; for the lower charts the most frequent words are in the centre, becoming less frequent moving out. For the bottom right, all the words are displayed horizontally; a proportion are written vertically in the others. Each does a reasonable job in highlighting some key words such as 'Iran' and 'deal', but they appear to contain a lot of less important words too, such as 'can' and 'will'.

---

[7]http://www.davidmccandless.com

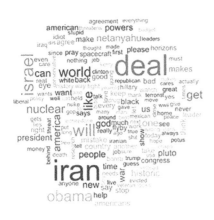

FIGURE 7.11 Word clouds showing the frequency of words appearing in tweets referencing various news channels: @BBCNews, @CNN, @FoxNews and @AJEnglish. Can you identify which is which?

7.6 DISPLAYING HOW DATA ARE DISTRIBUTED

Population pyramids

Figure 7.12 shows population pyramids for some of the ethnic groups enumerated in the 2011 Census population of England, as well as for the total population. Taking the top left pyramid as an example, on its left is what is essentially a bar

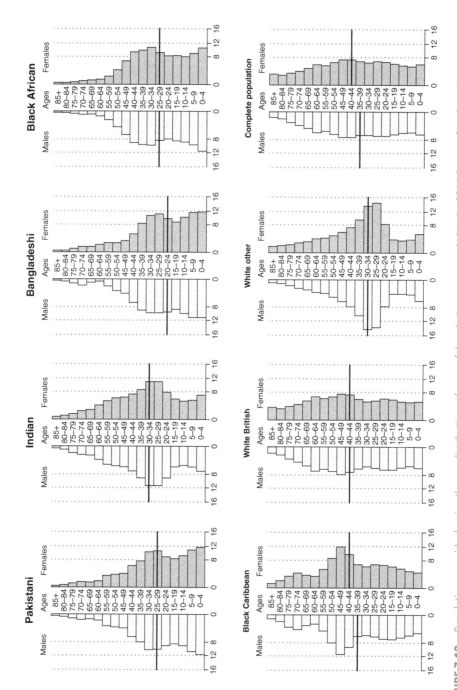

FIGURE 7.12 Population pyramids showing the age distributions of some of the ethnic groups recorded in the 2011 Census of England. The median age is indicated by the black line

plot, showing the percentage of the total male Pakistani population in each age category, and on its right the percentage of the female population. The median age for both male and female Pakistanis is the 25–29 age group. Generally, the minority groups are younger than the White British group, for whom the median was 40–44 years for both males and females.[8]

Although tailored to demographic data, the population pyramids are really just a means to show how the data (the ages of the populations) are distributed – how they spread out around their centre. This they have in common with the histograms and box plots below.

Strip charts, histograms, box and other plots

Figure 7.13 explores, for various European countries, the spread of electricity costs, expressed in euros per kilowatt-hour, with taxes and levies included. The simplest chart is the strip chart (Figure 7.13a), which places a symbol at the point on the number line that corresponds to the cost in each country. The prices range from €0.059 in Kosovo to €0.304 in Denmark. This chart suffers from overplotting – countries with equal or similar values obscuring each other – a problem that would worsen with more countries. A solution is to group the data a little, and this is done in Figure 7.13b by rounding the costs to two decimal places and then stacking duplicated values; it shows there are four countries with an electricity cost of approximately €0.12 per kilowatt-hour, and six with €0.13. A rug plot has been included – the tick marks at the bottom of the chart, which indicate the original, unrounded values.

For larger data sets, a stacked strip chart becomes unwieldy, so a histogram is used instead. For a histogram, the data are grouped into numerically consecutive groups (sometimes called bins) and the height of each bar is proportional to the number of observations in each group. A histogram looks a lot like a bar plot, but they aren't quite the same: the histogram is used for continuous data (measurements that could fall anywhere within a given numeric range); a bar plot is used for discrete data, to display the number of observations in each category (e.g. in each country). There is no gap between the bars in the histogram because there is no gap between the bins – the upper limit for one group is the lower limit for the next. The bars are ordered numerically by the values along the horizontal axis.

To give an indication of where in each bin the actual measurements fall, a rug plot is added. The histogram in Figure 7.13c implies the data range from €0.05 to €0.35 per kilowatt-hour; the rug plot shows the highest value is actually just above 0.30. A density estimate also is added to Figure 7.13c. We can understand this as showing the shape of the data without the 'blockiness' incurred by a histogram.

---

[8] 79.8 per cent of the usual resident English population was White British in 2011, although 44.9 per cent in London.

Figure 7.13d moves to a box plot, also known as a box and whisker plot. As well as showing the range, interquartile range, median and highlighting any potential out-liers in the data, box plots can be useful for comparing data – the prices for 2010 and 2014 are shown. Not surprisingly, the prices (which are not corrected for inflation) have risen, on median average, from 2010 to 2014. They also vary more between the countries. There are no obvious outliers in Figure 7.13d. If there were, they'd be shown as points positioned beyond the whiskers, the dashed horizontal lines.

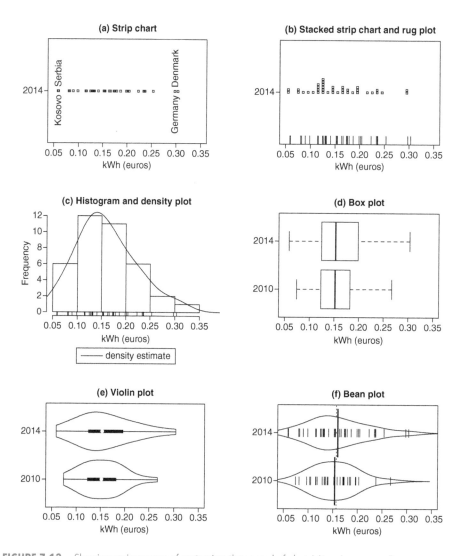

FIGURE 7.13 Showing various ways of portraying the spread of electricity prices across European countries (source of data: http://ec.europa.eu/eurostat/data/database)

The remaining graphs are similar to the box plots but provide greater visual information about the shape of the two data sets. The first is a violin plot, which is a box plot combined with a density estimate. The second is a bean plot, which is a rug plot combined with a density estimate with the median and the mean values shown too (Kampstra, n.d.). It would not be difficult to extend the bean plot to show, for example, the interquartile ranges. There is, however, a question of balance, presenting key information as clearly as possible without overcomplicating the graph and leaving it cluttered.

Which of the graphs is the best? Personally I prefer the bean plot, but it is not necessarily the easiest for a lay audience to comprehend. Keeping in mind the purpose and audience for a graphic is important. Many visualisations are not intended for publication but are used to help an analyst make sense of a data set – to identify any outliers or errors, for example. They form part of the process of interacting with the data during the analytical stages. Others are used to communicate findings from the analysis, which may not be to an expert audience. In both cases, it is important that the graphic does not mislead or misrepresent. However, in the second there are additional communication issues to consider. It is often said that a picture is worth a thousand words. Flipping that around, if it takes a lot of words to explain the picture, then it may be that its worth has been undermined.

7.6 REVEALING RELATIONSHIPS

Suppose we are interested in whether there is a correspondence between domestic electricity and gas prices in European countries. Is it the case that as one is higher so too is the other, or are higher electricity prices in countries such as Germany and Denmark offset by lower gas prices? To quantify any relationship between the two energy sources, there are various statistical methods we could use. We shall consider them in Chapters 9 and 10. For now, the aim is to explore the relationship visually, using a scatter plot.

Imagine a spreadsheet where each row represents a country and where there are two columns of paired data, one indicating the price of gas in 2014, the other the price of electricity (both in euros per kilowatt-hour). Those pairs can be plotted on a chart where the horizontal or X-axis corresponds to one variable and the vertical or Y-axis corresponds to the other. If high (above average) values of the X variable associate with high (above average) values of the Y variable, and low values of X with low values of Y, then it is a positive relationship in the sense that a line of best fit will be upward sloping (it will have a positive gradient). If high values of X associate with low values of Y, and low values of X with high values of Y, then the relationship is negative and downward sloping.

Figure 7.14 plots such data for 29 European countries in 2014. The data are the same in all four plots, but in each successive part of the figure we build up the

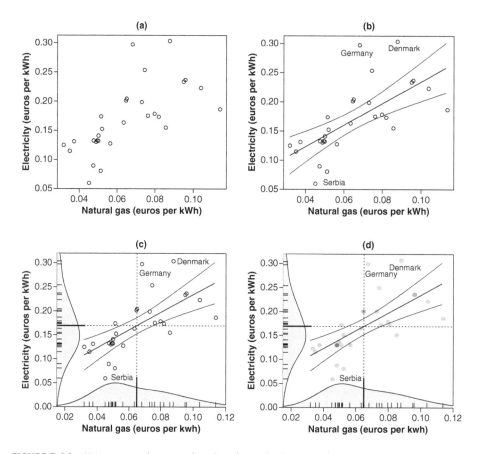

FIGURE 7.14 Using scatter plots to explore the relationship between electricity and gas prices in European countries in 2014: (a) is the basic scatter plot; (b) adds a regression line and the 95 per cent confidence interval around it; (c) includes part of a bean plot; (d) employs hexagonal binning (source of data: http://ec.europa.eu/eurostat/data/database)

amount of visual information. Figure 7.14a is just the basic scatter plot, and in Figure 7.14b a line of best fit is added. Lines of best fit can be misleading – they draw the eye and can disguise the fact that some data are either too noisy or too relationally complicated to be represented by a line. In this instance, however, it is generally the case that higher electricity prices are associated with higher gas prices (and lower prices with lower ones). It is not a perfect relationship by any means; there are outliers such as Germany and Denmark that are furthest from the line. Nevertheless, on average, as gas prices rise so too do electricity prices. Note that it is usual for the vertical, Y-axis to represent the variable that depends on the other. Hence Figure 7.14 implies that higher gas prices lead to higher electricity prices, which seems reasonable given that some electricity power stations are gas powered. However, it is unlikely to be a simple, direct and one-way relationship.

Figure 7.14b also adds a 95 per cent confidence interval around the line. Chapter 5 questioned the meaning of such an interval, especially in its traditional probabilistic interpretation that assumes random sampling of a population. Nevertheless, it provides a useful reminder of the uncertainty of the data, as well as helping to highlight some of the more unusual cases, those that are furthest from the line. Germany and Denmark have electricity prices that are high relative to the price of natural gas in those countries. In both, a part of this premium has been attributed to the transition to greener fuels. In contrast, in Serbia, electricity prices are low relative to gas.

Further annotation is provided in Figure 7.14c in the form of a bean plot. Note that where the peaks of the two distributions come together there is a cluster of points, some of which overplot each other. One way to address this is to cover the chart with a series of hexagons that are shaded in accordance to the number of data points that fall within them, ignoring those with a zero count. That is what Figure 7.14d provides.

Methods to deal with overplotting

In truth, overplotting is not much of an issue for what is a small data set with only 29 countries. For larger data sets it becomes more of a problem. Consider Figure 7.15. For Census zones across England and Wales, it explores the relationship between the percentage of households that do not own a car or van and the percentage of the economically active population that are unemployed. The ordering of the axes implies not owning a vehicle (the Y variable) is associated with being unemployed (the X variable). This is reasonable on the basis that affording a car/van is difficult if you do not have a job. Nevertheless, it would also be true that not having a vehicle increases the prospect of unemployment because jobs become harder to get to. In fact, it is possible that there is no relationship at all between unemployment and the lack of a car/van, despite what the graphs suggest. This is because the Census data are at the scale of neighbourhoods and not individuals: it is possible that those who are unemployed happen to live in areas with people who do not own cars but are not themselves those people. The assumption that a relationship found at one scale (the neighbourhood) must necessarily apply at another (individual people) is known as the 'ecological fallacy' and needs to be considered before reaching conclusions based on geographically aggregated data. It is discussed further in Chapter 9.

In any case, with 7201 observations, overplotting is much more evident. Figure 7.15a attempts the pragmatic solution, which is to reduce the size of the plotting symbol. This is only partly successful; the hexagons in Figure 7.15b do better. Other methods include adding transparency to the shading, so that where points

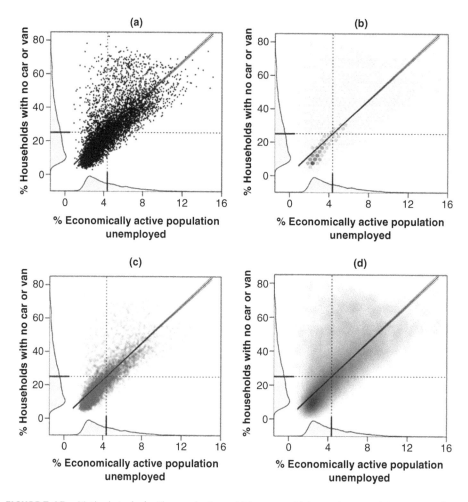

FIGURE 7.15 Methods to deal with overplotting, which occurs with larger data sets: (a) uses a smaller plotting symbol; (b) uses hexagonal binning; (c) uses transparency; (d) uses kernel density estimation

are densely concentrated the points appear darker (Figure 7.15c), and using methods of density estimation (in two dimensions) to indicate the density of points across the graph (Figure 7.15d).

Scatter plot matrices

The obvious limitation of a conventional scatter plot is that it is limited to two axes. What if there are three or more variables to consider? One option is to plot each variable against each other in turn, producing a matrix of scatter plots. Figure 7.16 is a scatter plot of the percentage of households exhibiting three or four dimensions

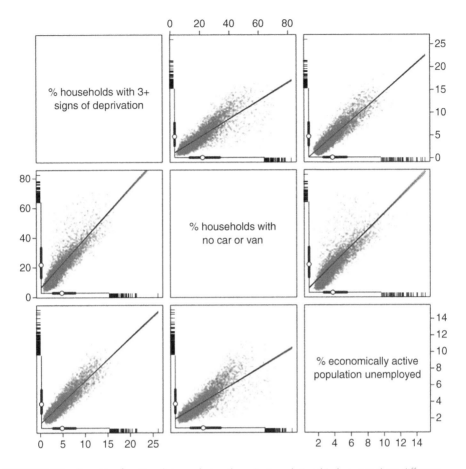

FIGURE 7.16 A matrix of scatter plots, exploring the pairwise relationship between three different variables

of deprivation (a UK Census classification of deprivation; four is the maximum), against the percentage of households with no car or van (top middle); the percentage of households exhibiting three or more dimensions of deprivation, against the economically active population unemployed (top right); and the percentage of households with no car or van, against the economically active population unemployed (middle right). The remaining set of plots is a mirror of the first set, with the X- and Y-axes reversed.

The matrix could be extended to include any number of variables, although the number of plots will increase quickly.[9] Because each scatter plot has only two

---

[9]The number of plots is $k^2 - k$, where k is the number of variables: four variables require 12 plots, 5 require 20, 6 require 30, and so forth.

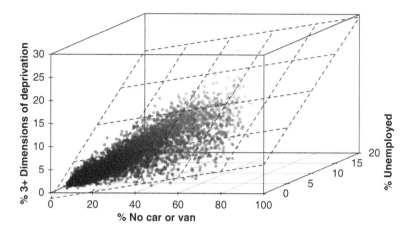

FIGURE 7.17   A three-dimensional scatter plot. Although it indicates the relationship between deprivation, car/van non-ownership and unemployment, it is not a straightforward graph to interpret

variables on it, a more fundamental weakness is that we are never looking at the relationship between the dimensions of deprivation, unemployment *and* car/van non-ownership simultaneously, just at the pairwise relationship between any two at a time.

3D scatter plot

One possibility to address this issue is to add a third dimension to the scatter plot, given there are three variables. In this case, and unlike the '3D' pie chart or other pseudo-3D effects, the use of a third dimension is justified: three variables, three dimensions. The problem is that the 3D plot is not easy to interpret. Looking at how the line (or, rather, plane) of best fit is angled in Figure 7.17, we can learn that the percentage of households with three or more dimensions of poverty tends to increase with the percentage owning neither a car nor van, and also increases with the percentage unemployed.[10] However, it is hard to read off the values for any one observation, to identify their exact location on the chart.

Differentiating by colour and symbols

A more elegant solution can be obtained by observing that although a traditional scatter plot has only two dimensions, it contains other visual components that can be changed to represent additional information: the shape, size and colour of the

---

[10]The relationship between unemployment and deprivation is not surprising: unemployment is one of the four dimensions of deprivation in the Census classification.

plotting symbols. Each one of these can be varied to indicate a value, a group or a category. This approach is used on the Gapminder website for its bubble charts showing, amongst other visualisations, carbon dioxide emissions by country since 1820: the X-axis is income per person (inflation adjusted), the Y-axis is CO_2 emissions per person; the size of the symbol is proportional to yearly CO_2 emissions, and it is shaded according to the geographical region of the country.[11] Additionally, the charts use animation to visualise change over time. You can produce your own motion charts as a graphic in Google Sheets.

Figure 7.18 is less ambitious, jointly using symbol size and colour to represent the third variable, the percentage of households exhibiting at least three of the four dimensions of deprivation. The radius of the circles is proportional to the

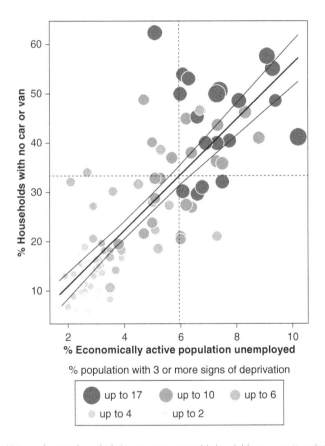

FIGURE 7.18   Using colour and symbol size to represent a third variable on a scatter plot

---

[11] Go to http://www.gapminder.org/world, open the graph menu and select from under the climate graphs.

square root of the percentage of households experiencing deprivation. Taking the square root means the area of the circle, A, is proportional to the levels of deprivation (because $A = \pi r^2$). Unsurprisingly, there are greater percentages of households experiencing three or four dimensions of deprivation in areas of higher unemployment and of non-ownership of cars and vans. However, the relationship isn't always that simple: to the top of the chart is an area with high deprivation levels but also below-the-mean levels of unemployment; to the bottom right are places with high deprivation but below-average levels of vehicle non-ownership.

Co-plots

Using different visual cues can be useful, but for a large data set it will be ineffective. Figure 7.18 takes only a small sample of the full data set ($n = 100$ of 7201) and yet there is already a problem with the circles overlapping and obscuring each other. The overplotting will prevent many of the observations being seen with a larger data set, thereby defeating the point of using colour or symbols to represent a third (or fourth) set of values.

An alternative approach is shown in Figure 7.19, which is for the full data set and shows the relationship between car/van non-ownership (Y) and unemployment (X) for different values of the third variable (Z), the percentage of households exhibiting three or four dimensions of deprivation. The chart begins at the bottom left and works rightwards and upwards: the scatter plot at the bottom left is the relationship between Y and X for $0 \leq z < 5$; the bottom right is for $5 \leq z < 10$; the top left is for $10 \leq z < 15$ and the top right is for $z \geq 15$. In other words, the bottom left is the relationship between car/van non-ownership and unemployment for the lowest percentages of household deprivation; the top right is the relationship for the highest. The chart makes clear that average car/van non-ownership and average unemployment increase with the percentage of households exhibiting three or more dimensions of deprivation, but there is variation around this general trend.

This type of chart is called a co-plot. It plots the relationship between two variables conditional on a third. At the moment there is one conditioning variable, but a second could be added. From a geographical point of view, an interesting possibility is to use the grid coordinates for where the data were collected as the conditioning variables (X = easting, Y = northing, where X and Y give the geographical centre of a Census area). The result, shown in Figure 7.20, is essentially a map, indicating how the relationship between car/van non-ownership and unemployment varies for different parts of England and Wales. London is in the box second from bottom and second from the right. The empty boxes are the sea.

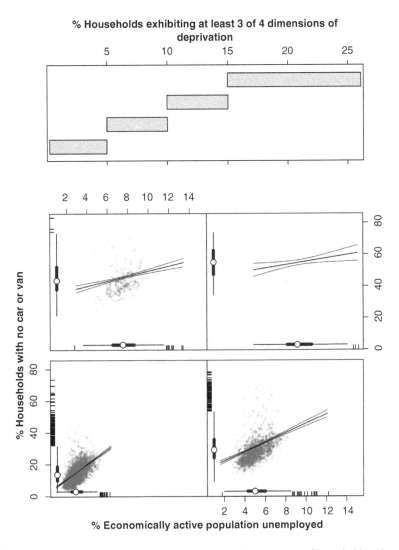

FIGURE 7.19 A co-plot exploring the relationship between the percentage of households with neither a car nor a van and the percentage of the economically active population unemployed, conditional on the percentage of households exhibiting three or four of the dimensions of deprivation

Parallel coordinates plot

Some further techniques for visualising the relationships between multivariate (multiple variable) data are shown in Figures 7.21 and 7.22.

The first of these is a parallel coordinates plot, which scales each variable from its lowest to its highest value, arranges the variables side by side and then connects

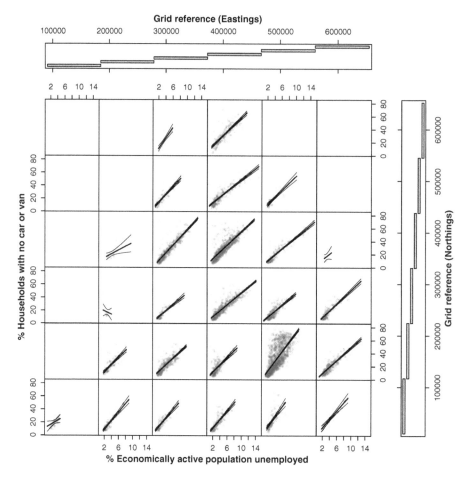

FIGURE 7.20 A co-plot with two conditioning variables, which, in this example, are geographical coordinates producing what is, in effect, a map indicating the geographically varying relationship between car/van non-ownership and unemployment

the values for each observation. In Figure 7.21 the fourth and fifth variables are the percentage of the adult population without a formally recognised educational qualification, and the percentage of those who declared a religion who said they were Christian. From the top left of the chart, and shown as a thick, dark line, there is a Census area where the percentage of households with three or more dimensions of deprivation is very high, car/van non-ownership is less high but above average, unemployment is high, there is a great lack of qualifications and a high percentage of Christians. It is likely that this represents an area of White, marginalised poor. A little below is an area with high rates of deprivation, of

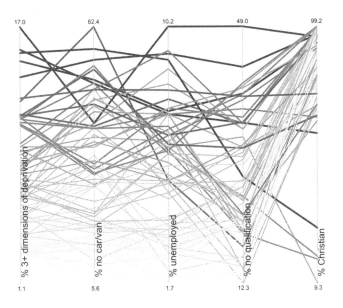

FIGURE 7.21   A parallel coordinates plot

car/van non-ownership and of unemployment but also lower rates of educational disadvantage and of Christians. This is likely to be an urban area, with a younger, ethnic minority population that is better qualified but still concentrated in areas of relative poverty. Again, Figure 7.21 only shows a small sample of the full data set. It is easy to imagine how it would become unreadable as the number of observations and therefore lines increased.

Radar plots

Finally, Figure 7.22 is a series of radar plots, also known as spider plots, where each plot represents a separate location. The shape of the shaded polygon (the web) is dictated by the amount of each variable measured at the location and, in this example, the dotted lines represent quartiles: the thicker, dashed line represents the median for each variable; the dotted lines working inwards from the medians are the first quartiles and minimums; the remaining dotted lines, working outwards, are the third quartiles and maximums. For example, the neighbourhood in Newport, Wales, has a percentage of Christians that is at the median for England and Wales, a percentage of the population lacking educational qualifications that is above the third quartile nationally, and similar for the remaining variables. The location in Shepway, Kent, is below the median on all variables except the percentage of the population that is Christian.

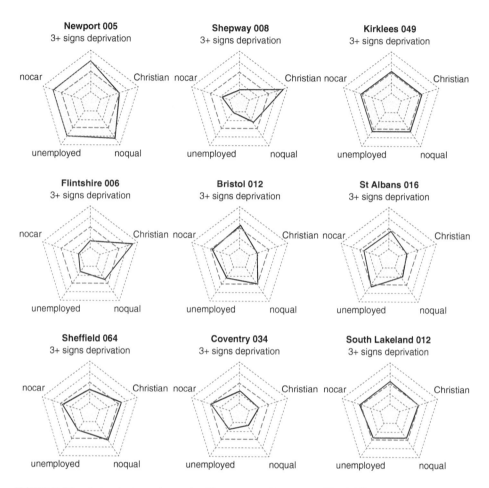

FIGURE 7.22   Radar charts can be used to illustrate a multivariate profile of different neighbourhoods and how they compare or differ

Radar plots have proved popular in geodemographic analysis – the like-with-like grouping of residential locations into a relatively small number of neighbourhood types – where the plots are used to portray the characteristics and differences between the types (Harris et al., 2005). However, one expert on graphical design has them on his list of 'silly graphs that are best forsaken', which takes us full circle – his verdict is the same as that against pie and doughnut charts: the information would be clearer in a bar chart (Few, 2012).[12]

_____

[12]I am not sure I agree: the bar plot may be better where there are relatively few variables to portray, but a radar plot can pack a lot of variables into a small graphical space in a way that a bar plot would struggle to achieve.

7.7 CONCLUSIONS

There are many more types of chart and graphic than appear in this chapter and new ones are created regularly, in part by people's imagination and creativity, in part because new technologies enable their design, and in part because new forms of data can require new forms of presentation. The graphics in this chapter are, of course, static, but by using online or other computer-enhanced visualisations, it is possible to interact with data and to query it 'on the fly', as well as to create linked data presentations that allow an observation to be identified in one chart and at the same time highlighted in graphs and tables. At the time of writing, there is an interesting collection of interactive graphics from the *New York Times* and the *Guardian* at http://collection.marijerooze.nl, which includes 'Summer love your best or worst summer dates', 'Obama's health law: who was helped most' and 'Flooding risk from climate change, country by country'. Tools such as Google Fusion Tables, Google Spreadsheets, Google Charts and Tableau Public can be used to create online data visualisations.

All the graphics in the chapter were created in the open source software, R, a widely used software environment for statistical computing and graphics. It is available for free from https://www.r-project.org/. A considerable advantage of this software is that the user can build upon the base graphics libraries with those that have been made available by other people to create population pyramids, undertake hexagonal binning or to produce radar charts, for example.[13] Moreover, it is relatively straightforward to then customise those charts. In this way, there is more control over the design process and no need to accept the often unsatisfactory default settings that are found in many other software packages. Plus, when your publisher reads a draft of this chapter and recommends a few changes, it takes only a few minutes to tweak the scripts for the graphics and reproduce them rather than spend a frustrating few hours or days trying to remember exactly which buttons you need to press or drop-down menus to select to recreate what you did in the first place. R takes a little getting used to but is worth the effort. The user manual, *An Introduction to R* (R Core Team, 1999–2015), can help, as well as books such as Winston Chang's *R Graphics Cookbook* (2013).[14] There is a brief introduction in Chapter 12.

---

[13] The packages used for this chapter are: RColorBrewer, to generate colour palettes; twitteR, to interact with Twitter; wordcloud, to produce word clouds; vioplot, for the violin plots; beanplot, for the bean plots; fMultivar, for the hexagonal binning; scatterplot3d, for 3D scatter plots; MASS, for the parallel coordinates plot; fmsb, for the radar charts; and code, by Markus Gesmann at https://gist.github.com/mages/5019772, to produce the doughnut plots. These graphical methods may also be available in other R packages.

[14] The manual is free with the software or can be downloaded from https://cran.r-project.org/

In the final analysis, data presentation is as much an art as a science, mixing the quantitative with qualitative elements of design, aesthetics and what solicits a positive response in the reader. There will be difference of opinion on what makes for the best graphics, given an element of personal taste and because different designs serve different purposes. The *State of the Nation* report (British Future, 2013) contains graphics that violate many of the principles of a more traditional approach to data presentation, but it is more visually attractive as a consequence.

Nevertheless, there are better and worse diagrams. To see a contender for worst, do a web search for a 'Venn pie-agram' (although I am not yet persuaded that it isn't a hoax).[15] For your own reports and assignments, you may wish to consider the following suggestions. First, although it may make sense to highlight particular aspects of the data, it is dishonest to use a graphic to deliberately distort or mislead. Second, don't be reliant on a software package's default setting: a few tweaks and changes will often make for a better graph. Third, if you are using colour, choose colours that will reproduce well on a black and white printer, and also ones that will be discernible to people with a visual impairment such as colour blindness (red and green combinations especially are best avoided). The ColorBrewer website is helpful here, offering guidance on effective colour schemes.[16] Fourth, don't use novelty graphics for the sake of it, including those that add a pseudo-third dimension for no good reason. Fifth, don't overcomplicate the graphic unnecessarily. Sixth, and finally, exploding pies are an unfortunate culinary experience, not a sensible method of data presentation.

---

[15]http://flowingdata.com/2012/05/23/venn-pie-agrams/
[16]http://colorbrewer2.org

8

MAPPING AND GIS

8.1 INTRODUCTION

The previous chapter looked at ways of presenting data but it neglected something important to geography's cartographic traditions: the mapping of geographic data. This chapter makes amends for that omission, concentrating on some of the principles behind Geographic Information Systems (GIS) – software for the storage, analysis and visualisation of geographic information.

Figure 8.1 sets the scene. It is a stylised map of London's boroughs, showing the ethnic composition of their resident population according to the 2011 Census. It would be possible to present these data in a table, and it would be better to do so if the exact population numbers were required. However, the map reveals something that the table would not: the ethnic geography of London – its geographical patterning. Despite reports about the declining number of people in London who regard themselves as White British, this group remains dominant in the capital. There are clusters of people of Asian Commonwealth heritage (Bangladeshi, Indian or Pakistani), especially in the west, northwest and east of London. Black groups (Black African, Black Caribbean or Black other) are found especially, but not exclusively, to the southeast of the centre. Most of the boroughs contain an ethnically diverse population, although that is less true of some at the outer boundary of London.

The map is a simplification of a complex reality. It is a cartographic product – a pictorial abstraction – but it is still linked to data that are the outcome of social, economic and demographic processes that are actually happening on the ground. Knowing where something occurs can be a first step towards understanding why it occurs or helping to manage its impact. GIS are technologies that help manage, make sense of and draw meaning from geographic data.

8.2 VIEWS OF THE WORLD

Figure 8.2 gives two different maps of the 2001 Census population of Northern Ireland. The first is a raster view; the second is vector. The differences between the

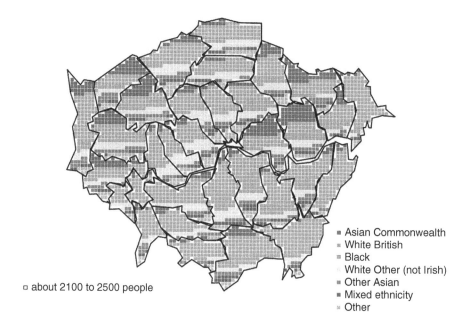

about 2100 to 2500 people

- Asian Commonwealth
- White British
- Black
- White Other (not Irish)
- Other Asian
- Mixed ethnicity
- Other

FIGURE 8.1 A map showing the ethnic composition of London's boroughs in 2011 (source of data: UK Census)

maps are more apparent in Figure 8.3, which zooms into the area in and around Belfast city centre. The blocky, grid structure of the raster map contrasts with the polygonal structure of the vector map.

Raster and vector are the primary data models used in GIS. These are the sorts of desktop software used to store, query, manipulate, analyse and display geographic data. Geographic data, also known as spatial data, can be mapped because they include a georeference identifying where an observation (a measurement) has been taken – a grid coordinate, or a longitude and latitude, for example. Geographic data combine what has been measured (the attribute data) with where it has been measured (the location or georeference). There are a number of well-known commercial GIS, and also free software including the popular QGIS package, available for Windows, Mac and Linux.[1] The content of this chapter was produced in R, the open source, statistical computing and graphical software mentioned at the end of the previous chapter. Its various spatial analytical libraries provide a powerful open source tool for geographical analysis and mapping.[2] Chris Brunsdon's and Lex Comber's book *An Introduction to R for Spatial Analysis & Mapping* (2015) offers what its title suggests; Bivand et al.'s *Applied Spatial Data Analysis with R* (2013) is a little more advanced.

---

[1]http://qgis.org/en/site/
[2]https://cran.r-project.org/web/views/Spatial.html

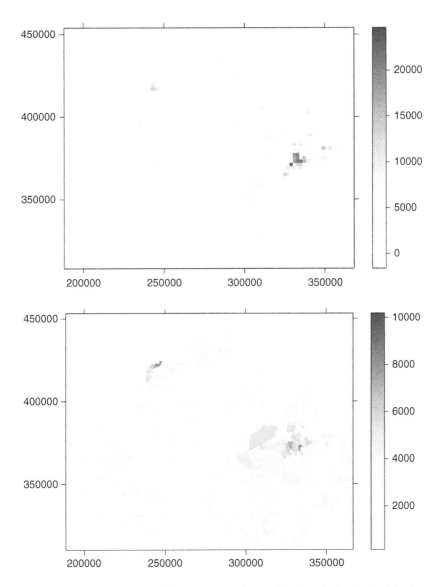

FIGURE 8.2 A raster representation of the Census population of Northern Ireland in 2001 (top). A vector representation (bottom) (source of data: http//www.nisra.gov.uk/geography/home.htm)

8.3 RASTER DATA

A raster is a grid structure. At its simplest, it is a series of values, one for each of the grid squares, also known as cells. The Northern Ireland map is a 73 row by 90 column grid, the product of which is 6570 cells and therefore 6570 values.

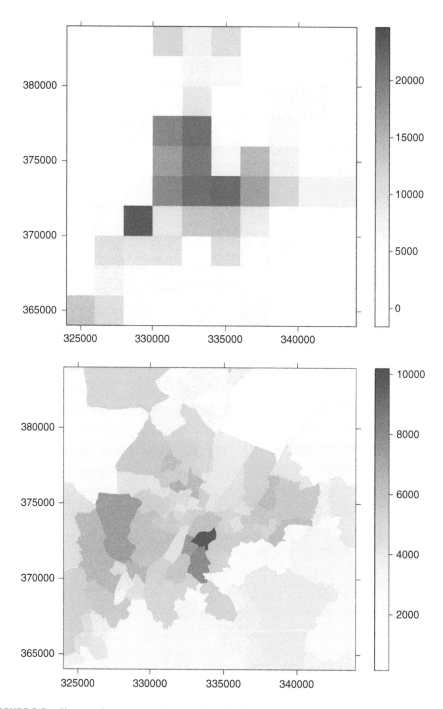

FIGURE 8.3 Showing the areas in and around the city of Belfast: in raster format (top); in vector format (bottom)

Not all of those cells record a population count, because they are offshore and in the sea. They could be assigned zero population, but it is better to say they contain no measurement at all. They will then appear unshaded in the maps and not distort any statistical calculation. There is an average of 444 persons per grid cell on the land, but that average drops to 257 if the offshore cells are taken as zero, giving a misleading indication of the population density. Therefore, in addition to the measured values, an NA (not applicable) value is required to indicate an empty cell.[3]

To map the data, the locational information is required. The geography of the grid is defined by assigning a geographical coordinate to one corner and combining that with knowledge of the number of rows (*nrow*), the number of columns (*ncol*), and the length/width of each grid cell. In addition, it may be necessary to include a coordinate reference system (CRS), especially for a raster covering a large area of the Earth's curved surface. The CRS will identify the ellipsoid used to model the shape of the Earth, the datum defining the origin ((0,0) point) of the coordinate system, and the projection required to flatten the globe and draw it on a map.

The area of each grid cell is the cell size and defines the resolution of the grid. The raster shown in Figures 8.2 and 8.3 is of size 2 km × 2 km, or 4 km². Decreasing the cell size increases the resolution of a grid, decreasing the 'blockiness' of the raster and the appearance of pixelation. However, it increases the size of the file and therefore the computational demands. Halving the cell size halves the length and also the width of each cell, quadrupling the number of cells. If each value requires 2–8 bytes of memory, then high-resolution rasters covering a large study area will quickly become voluminous. Figure 8.4 is a digital elevation model (DEM) showing the height of the Rocky Mountains from just north of the Canadian–US border (top left of the image), down to New Mexico. The heights are represented by a grid of 8.64 million cells, with a file size of about 17.3 MB. It can also be downloaded as a grid of over 138 million cells and a file size of 277 MB.[4] In either case it is only a small extract from the Global Multi-resolution Terrain Elevation Data 2010 (GMTED2010).

The raster values can be stored in an *nrow* by *ncol* matrix. However, where there are many zero or, alternatively, many NA values, the matrix becomes sparse of other numbers. There may then be more efficient ways of storing the data, including methods of data compression. Some methods of compression are lossless, in that

---

[3]The same arguments could be extended to Lough Neagh to the west of Belfast and to Lower and Upper Lough Erne near the western borders with the Republic of Ireland, indeed to any 'empty' land including the various mountain ranges. How we classify the data will affect the analysis that follows on from that classification; uncertainty in the classification will create uncertainty in the results.

[4]From http://earthexplorer.usgs.gov/

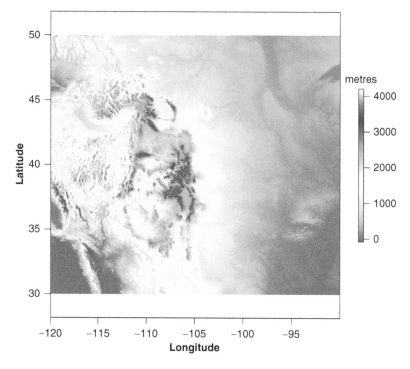

FIGURE 8.4 Digital elevation model showing the height of the Rocky Mountains through the United States (source: GMTED2010)

they reduce the file size but the process can be reversed to recover the original values. Others are lossy, in that they simplify but also degrade the image quality. The image file format, TIFF, may be used to store raster images, with or without compression.[5] A GeoTIFF allows a CRS to be embedded in the TIFF.

8.4 EXAMPLES OF RASTER ANALYSIS

Cell aggregation

Figure 8.5 shows some typical raster processes. The first is a simple aggregation of the data – in this case, the 2 km × 2 km grid cells of Northern Ireland's population are grouped into larger, 20 km × 20 km squares. The length of the larger squares is ten times the original grid cells, giving 10 × 10 grid cells per new square. Summing together the values for those 100 cells gives the total population count for the larger cell. Other functions could be applied during the aggregation, to find the minimum, maximum, mean or median values, for example.

---

[5]JPEG files usually are smaller, but employ lossy compression.

Cell reclassification

Figure 8.5b illustrates an even simpler raster operation: a reclassification of the original grid values. Suppose all cells with a population count of 400 or greater are defined as built-up areas/urban settlements. The raster can be then reclassified such that $z < 400 \rightarrow 0$ and $z \geq 400 \rightarrow 1$, where z is the original raster value.

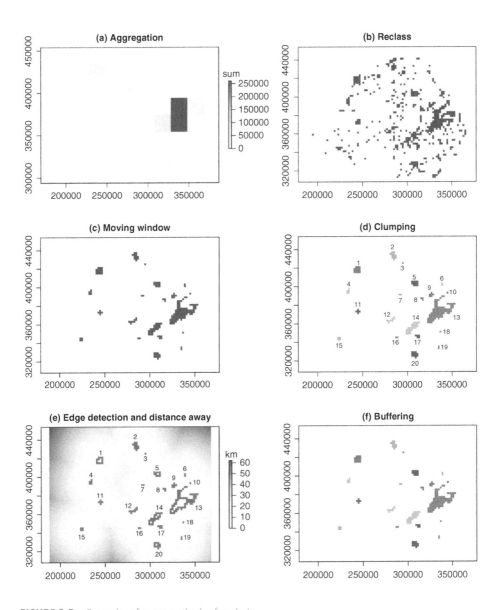

FIGURE 8.5 Examples of raster methods of analysis

The data are now categorical: the values of 0 and 1 have no quantitative meaning other than to distinguish one category (low population density/rural) from the other (higher population density/urban). Sometimes categorical data are described as qualitative because the numbers are arbitrary and simply a substitute for words (0 = rural, 1 = urban). Personally, I try and avoid it: the data are not qualitative in the sense that many in the social sciences and humanities would mean it, where focus groups, discourse analysis, ethnography, participant observation and other non-quantitative approaches are adopted.

Moving windows

Figure 8.5c illustrates a moving window operation, also known as a focal operation because it focuses (centres) on each grid cell in turn. Figure 8.6 elaborates upon the process. In this example, a 3 × 3 moving window is centred upon one of the grid cells. The number at the centre is then substituted in a new grid with the median of the nine values within the window of the old. The process is repeated, centring the window on each of the other cells in turn. The result is one of spatial smoothing – it removes the speckle in the map. Increasing the size of the window increases the amount of smoothing. Functions other than the median may be employed. Methods similar to moving windows are used in creating geographically weighted statistics (see Chapter 11). A potential problem is one of edge effects. Along any of the sides, the window extends beyond the study region, with a maximum of six cells contained within it and four at the corners.

Clumping

Figure 8.5d shows the results of a clumping operation, detecting patches of connected cells. Connections can be defined in two ways: the rook's or queen's case. By analogy with chess, where the rook can move forwards, backwards, from side to side, but not diagonally, the rook's case connects only cells that meet along a

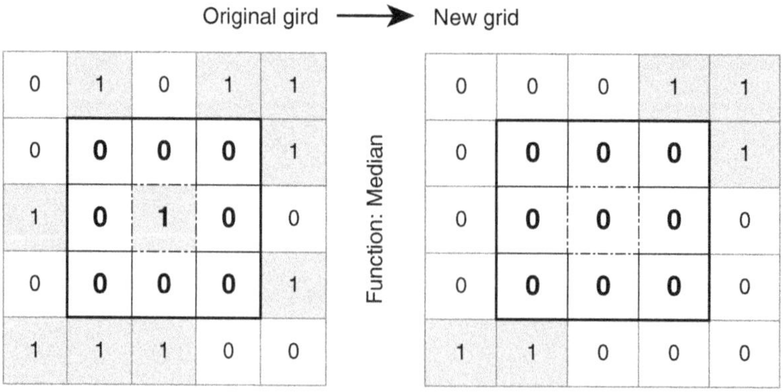

FIGURE 8.6 Illustrating a moving window operation

shared boundary (the shared side of two cells). The queen can move diagonally, so the queen's case also connects cells that meet at a corner.

Edge detection, distance calculations and buffering

Figure 8.5e arises from a process of edge detection. Finding an edge is a moving window operation: an edge exists where not all the values are the same and therefore both the range and standard deviation are greater than zero. Also shown is the distance of cells to the nearest clump of higher population density. From those distances, a 10 km buffer region is drawn around the settlements in Figure 8.5f.

Map algebra

Map algebra is another common raster operation. If two grids are geographically aligned and have the same resolution then the cell values for their corresponding locations can be added, subtracted, multiplied and divided in a straightforward way. Figure 8.7 provides an example.

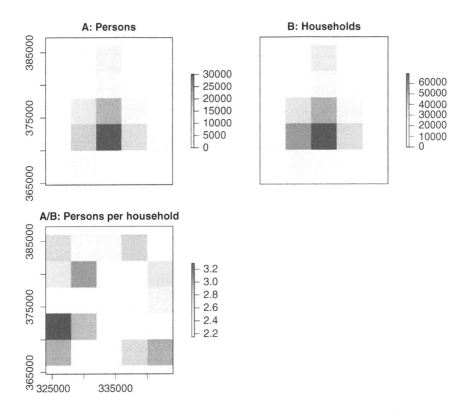

FIGURE 8.7 An example of map algebra. In this case, the number of persons per household is calculated by dividing the raster count of population by the raster count of households

Route analysis

Figure 8.8 show the result of a more sophisticated raster procedure, determining the least-cost route from A to B, where the cost increases with the height difference in traversing from one cell to the next. In other words, it seeks out the flattest route, balancing distance with the total ascent and descent.

Together, these and other raster processes provide a toolkit for location analysis, risk assessment, terrain analysis, measuring and modelling land use and handling remotely sensed data from aerial sensor and satellites, amongst other applications.

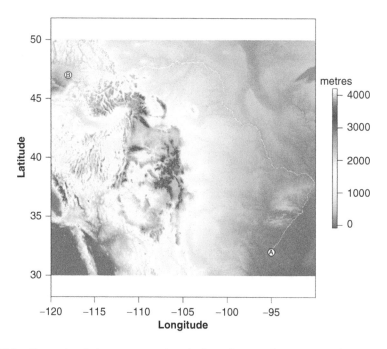

FIGURE 8.8   The results of a least-cost procedure, finding a flat route from A to B on the map

8.5 VECTOR DATA

The vector data model is more complex than raster's grid structure, although it reduces to a simple idea: locating a point on a map. Vector data come in three formats: lists of points, lists of lines and lists of polygons. A point is defined by a geographical coordinate, (x, y). A straight line has a start point, (x_1, y_1), and an endpoint, (x_2, y_2). A series of lines, plotting the course of a river for example, can be represented by a 'polyline', joining multiple points together: $\{(x_1, y_1), (x_2, y_2), (x_3, y_3), \ldots, (x_n, y_n)\}$. Closed shapes (i.e. polygons) are produced when the start and the end points are the same: $\{(x_1, y_1), (x_2, y_2), (x_3, y_3), \ldots, (x_n, y_n), (x_1, y_1)\}$.

Points, lines and polygons represent geographical features on a map. For example, the location of a shopping mall, the path of a railway or the boundary of a city. In addition to this geographic information, it is common to have some information about the features' attributes – their name, their length, the number of people who live there, and so forth. This is usually stored separately from the geographic information, with a shared ID to link them back together. This link allows features to be searched for by attribute and to be highlighted on the map; reciprocally, it allows for features to be selected on the map and the information about them to be displayed in the attribute table.

Efforts are made to avoid redundant information. Imagine there are two neighbourhoods and that they share a boundary along one side. The perimeter of neighbourhood A could be recorded as one list of point coordinates; the perimeter of neighbourhood B as another. However, the two separate lists will duplicate the information for the point coordinates that lie along their shared border. A more efficient system is to create a series of tables. First, a table listing all the points for all the features in the study region, with each row identifying a point and its coordinates (without duplication). Second, a table of line segments, with each row identifying where in the points table the geographical coordinates of the lines' start and end points can be found. Third, a table of polygons, identifying which line segments form the polygon. These tables cascade downwards: the polygons data link to the lines data that link to the points data. In addition, topological information may be recorded such as which polygons lie on either side of a line. Topological information is used to detect and to ratify geographical errors in the data (e.g. unclosed polygons). It can also speed up some analytical operations such as deleting the boundary between neighbouring polygons to merge them.

As with raster data, a coordinate reference system may be required to map the data correctly. Both R and QGIS utilise EPSG codes from an online data set and registry that can be queried at http://www.epsg.org. Searching on code 3857, for example, provides information about the coordinate system used by Google Maps.

Figure 8.9 shows three layers of vector data where the points (representing trees) are overlaid upon the lines (representing paths and highways) that, in turn, are overlaid upon the polygons (representing buildings). The region is the area around the Brandenburg Gate, Berlin. The data are from OpenStreetMap, which is 'an initiative to create and provide free geographic data, such as street maps, to anyone.'[6] You can download the data or contribute to the map at http://www.openstreetmap.org.

Vector data typically are available in shapefile format (.shp). A shapefile is actually a collection of files that are stored together and have the same name but different extensions: for example, mymap.shp, mymap.shx, mymap.dbf, mymap.prj.

---

[6]http://wiki.osmfoundation.org/wiki/Main_Page

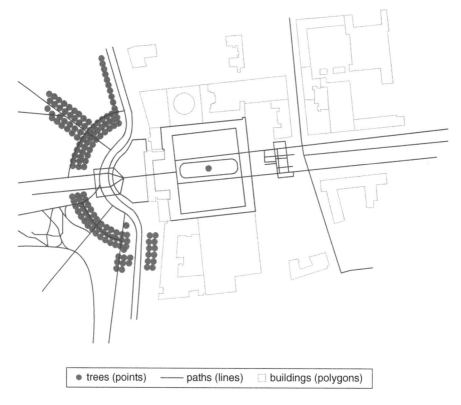

| ● trees (points) | ——— paths (lines) | ☐ buildings (polygons) |

FIGURE 8.9 Three layers of vector data showing the area around the Brandenburg Gate (source of mapping data: OpenStreetMap)

The first two of these are for the locational information and provide the geography for the map; the third is for the attribute data. The fourth, which is optional, contains the CRS.

8.6 EXAMPLE OF VECTOR ANALYSIS

Vector data are well suited for geometric operations, a number of which are shown in Figure 8.10. Within England, schools do not have fixed catchment areas, but their *de facto* catchments can be modelled by mapping the locations of pupils who go to a school. These are what the points in Figure 8.10a represent. The school is represented by the + symbol, and the task is to define the core catchment area around it. One possibility would be to draw a circle around the school, extending the catchment out to a given radius. However, to do so assumes that the pupils are spread out evenly around the school, which, looking at the map, is not actually the case.

The bounding box and convex hull

The bounding box is the minimum-enclosing rectangle that contains all the points within the study region. It is shown as a dotted line in Figure 8.10a. It exaggerates the spatial extent of the school's catchment by including a lot of empty space where no pupils attending the school live. Shown as a solid line, the convex hull does better. A convex hull defines the smallest polygon that contains all the points, where all the interior angles of the polygon are less than 180 degrees. Although smaller than the bounding box, it still includes spatial outliers – the pupils who live at greatest distance from the school and remote from other pupils. A better model of the school's core catchment would exclude those outliers. One way to do so is to use a distance calculation: the distance of each pupil from the school, excluding those who are furthest away. However, this is equivalent to drawing a circle around the school and omitting those beyond a threshold distance. Again, it wrongly assumes an isotropic plain where the pupils are equally distributed in all directions around the school.

Voronoi polygons and Delaunay triangles

What matters is not just how far pupils are from the school but also how far they are from other pupils attending the same school.[7] Voronoi, also known as Thiessen polygons, and their close relation, Delaunay triangles, are methods of space-filling from point data. All the space within the boundary of a Voronoi polygon is closer to the point around which the polygon has been drawn than it is to any other of the points. In this way, the polygon defines a point's area of influence. Delaunay triangles can be used to model three-dimensional data, to create terrain models from irregularly spaced point data, for example.[8]

Figure 8.10b illustrates Voronoi polygons (and also Delaunay triangles) created for the whole of the study region but with only an area around the school shown, to allow it to be seen more clearly. The points are the pupil locations, the solid black lines the boundaries of the polygons, the dotted lines the triangles. Looking at the image, it may be observed that the further a pupil is from their nearest neighbours, the larger the polygon around them. Therefore, the spatial outliers are those for which the Voronoi polygons are largest.

Choosing subsets by attribute

By storing the area of each Voronoi polygon in an attribute table, the largest can be searched for and omitted. Doing so leaves a subset of the polygons, representing the part of the study region where the points and therefore the pupils are closest together. This is shown in Figure 8.10c.

---

[7] Harris et al. (2016) use an algorithm that includes both.
[8] They create TINs: triangulated irregular networks.

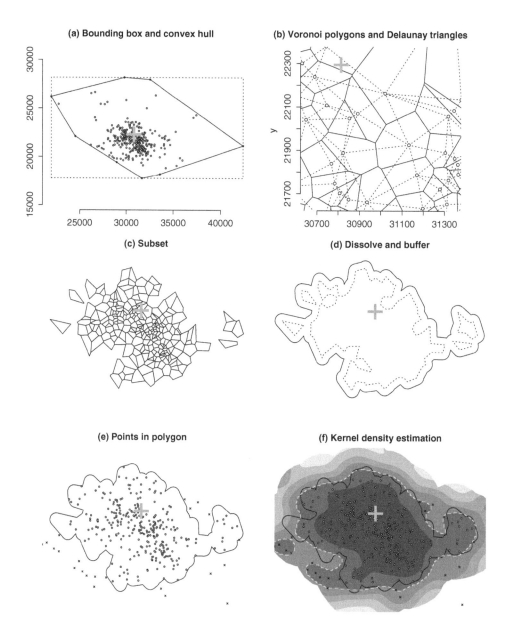

FIGURE 8.10 Examples of vector analysis, used to model the core catchment area of a school (see text for details)

Dissolving and buffering

The dissolve operation, shown in Figure 8.10d, removes the boundaries between polygons. Having done so, the catchment area consists of five disconnected parts – the large central area and four 'islands' around it. These can be joined into a single whole by using a buffering process.

Points in polygon

A common vector procedure identifies which points fall within which polygons. This is an overlay operation and can be used to assign the attributes of the polygons to the points they contain (e.g. each point can be given the ID or name of the polygon that contains it). In Figure 8.10e the process is used to identify which of the pupils are within the modelled catchment area and which are not. The method can be extended to consider whether the social and demographic attributes of pupils who are in the catchment area and who go to the school are similar to those of the pupils who are in the catchment area but do not go to the school. This approach can be used to look at processes of social sorting in the admission processes to schools: do some schools disproportionately attract the highest-attaining pupils? Is there evidence that different ethnic groups are attending different schools from one another?[9]

Kernel density estimation

Figure 8.10f confirms that, in general, the modelled catchment area corresponds to the areas with the highest density of pupils, as indicated by the shading. Kernel density estimation (KDE) was used as the basis for the shading. This procedure is similar to that used to generate density estimates of how a variable is distributed (see Chapter 7), except that the distribution is now in a two-dimensional geographical space and employs a moving-window type method to identify spatial clusters of points. The KDE is a method of rasterising the vector data: its input is a list of vector points; its output is a grid of density values. The KDE could itself be used to model the catchment area – for example, as the area shown within the white dotted line.[10]

Unions, intersects and clipping

Finally, Figure 8.11 illustrates an instance of two overlapping polygons representing two school catchment areas, for school A and for school B. The total area served by the schools is found by a union operation, combining their catchments. Also shown is the area where the schools most obviously compete for pupils, which is where their catchment areas intersect. A related operation (not shown) is to clip one vector layer to the spatial extent of another. We might, for example, have a polyline file showing all the interstate highways in the United States, and a

---

[9]See, for example, Harris and Johnston (2008).

[10]KDEs are used in conjunction with a space-filling method to remove gaps in the modelled catchment area by Singleton et al. (2010).

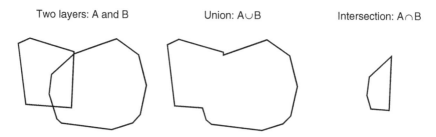

Two layers: A and B   Union: A∪B   Intersection: A∩B

FIGURE 8.11   The union and intersection of two vector layers

polygon file giving the boundary of Florida. Clipping the first to the second leaves a map of the highways traversing the Sunshine State.

8.7 VECTOR OR RASTER?

To reiterate, vector and raster are the two main data models used in GIS. In general, vector data are used to represent discrete objects: those that have a clear boundary, route or point location. They include Census neighbourhoods. However, the boundaries can be somewhat forced – designed for administrative convenience rather than to reflect the 'natural' boundaries of communities or other population distributions. Diversity within the areas can be concealed. Rasters are often used to represent more continuous distributions that don't stop at clear boundaries – for example, the height of an undulating surface. The analysis of raster data can be faster because of its simple grid structure, but, as noted, the file sizes can be large (although not necessarily greater than vector). Vector data is usually easier on the eye. The lack of pixelation makes for more aesthetically pleasing maps.

To some extent, the formats are interchangeable: raster can be vectorised; vectors can be rasterised. Raster cells can be used to represent points, lines or polygons. Elevation and other surfaces can be represented as contour lines on a map, and grid structures can be replicated as a series of polygons (as squares, known as vector grids). There is an old adage, 'vector is better, but raster is faster', but the truth of that statement depends on the context: ultimately the choice will be governed by how the data are collected and the application for which they will be used.

A third data model is a network structure showing the connections and interactions between people or places. A spatial weights matrix can be represented in this way, as a connectivity graph (see Chapter 11).

8.8 HOW TO LIE WITH MAPS

In a series of books that includes *How to Lie with Maps* (1996), *Air Apparent: How Meteorologists Learned to Map, Predict, and Dramatize Weather* (1999), *Bushmanders and*

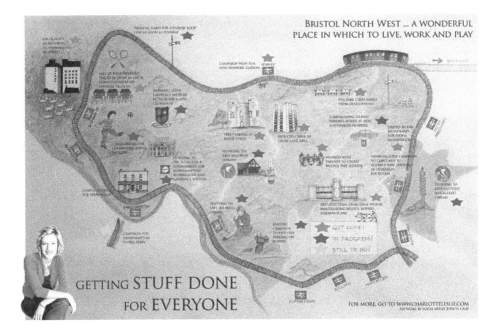

FIGURE 8.12 A map designed to support political campaigning (source: http://charlotteleslie.com/getting-stuff-done-for-everyone-our-amazing-map-of-bristol-north-west/)

Bullwinkles: How Politicians Manipulate Electronic Maps and Census Data to Win Elections (2001), and *No Dig, No Fly, No Go: How Maps Restrict and Control* (2010), Mark Monmonier explores the persuasive power of maps and how they are used and abused. Similarly, in *The Power of Maps* (1993) and *Rethinking the Power of Maps* (2010), Denis Wood shows how maps do not merely present things as they are but are instruments of communication, persuasion and power that reflect their authors' interests and points of view.

What we see in a map is a function of how the map is designed, and for that there are choices: how to represent real-world features, which to select, which to omit, whether to group them in some way, which colours to use, and so forth. Figure 8.12 is a clear example of a map being used for political purposes. It is from the website of the Member of Parliament for Bristol North West, near where I live. I have no objection to the map; it is a creative way of presenting the MP's involvement in local issues. However, it is certainly not impartial. I imagine voters for other political parties would present the map in very different ways.

Map classes

Figure 8.13 is more prosaic but makes clear how different choices affect the appearance of a map and therefore how it is read. In England, 80 per cent of the population classified itself as White British from the categories available in the 2011 Census. That leaves one-fifth from other groups that include Black Africans, Black Caribbeans, Bangladeshis, Indians, Pakistanis, Chinese, other White groups and those of mixed ethnicity. Pick a place at random, though, and you are unlikely to find that 20 per cent of its population is from a group other than the White British. The reason is that groups are not evenly spread across England but concentrated in particular urban locations (although less concentrated than they have been in the recent past).

Where are these locations? Using Figure 8.13a as a guide, you'd probably conclude that they are mainly in London. It is a choropleth map where each area – each local authority – is shaded according to the percentage of its resident population that is not White British. Specifically, it is shaded according to which of four groups its percentage falls into. Those groups, which are the map classes shown in the legend, split the data into four equally spaced intervals.

Whilst the map classes are equally spaced, the same cannot be said of the data. Those are unevenly spread between the lowest and highest values. There are far more observations in the bottom category (2.4 to less than 22.62 per cent, for which there are 261) than there are in the top category (63.05 to 83.27 per cent, 11 observations). Consequently, most of the map has the lightest shade and little has the darkest. In contrast, Figure 8.13b shades the map so that approximately the same amount of the map is filled by each colour. This has the effect of greatly expanding the top category, which now ranges from 11.49 to 83.27 per cent and contains more observations than any other (133). This happens because the local authorities containing the highest percentages of groups not White British are also those of smallest area (but highest population density), requiring more of them to be grouped together to fill one quarter of the map.

Figure 8.13c is a quantile classification – specifically, a quartile classification, given there are four groups. It aims to have the same number of observations in each group. The not White British groups do not appear as dispersed as they did in Figure 8.12b but still more so than in Figure 8.12a, giving greater emphasis to the Midlands and the North West of England.

In Figure 8.13d I have chosen a manual classification, which I have set at the 50th, 75th, 90th and 100th percentiles of the distribution. Like the equal interval classification, this has the effect of emphasising the places where the percentages of the population not White British are highest. The distribution of the percentage not White British variable is shown in the strip charts of Figure 8.14, which also show how the various map classes dissect the data. An alternative classification, found in some GIS packages, is what is called a natural breaks classification. What it tries to do is choose thresholds that reflect the shape of the distribution. For the current data, it produces breaks at 11.75, 26.08 and 47.83 per cent, as well as at the minimum and maximum values of 2.40 and 83.27: five breaks in total, to give four map classes.[11] If we change the number of map classes, then the breaks will change too. That will further affect the appearance of the map.

---

[11]The number of break points is always the number of map classes plus one.

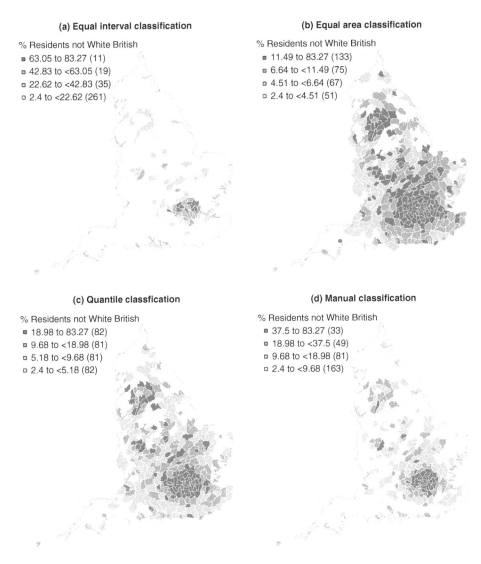

(a) Equal interval classification

% Residents not White British
- 63.05 to 83.27 (11)
- 42.83 to <63.05 (19)
- 22.62 to <42.83 (35)
- 2.4 to <22.62 (261)

(b) Equal area classification

% Residents not White British
- 11.49 to 83.27 (133)
- 6.64 to <11.49 (75)
- 4.51 to <6.64 (67)
- 2.4 to <4.51 (51)

(c) Quantile classfication

% Residents not White British
- 18.98 to 83.27 (82)
- 9.68 to <18.98 (81)
- 5.18 to <9.68 (81)
- 2.4 to <5.18 (82)

(d) Manual classification

% Residents not White British
- 37.5 to 83.27 (33)
- 18.98 to <37.5 (49)
- 9.68 to <18.98 (81)
- 2.4 to <9.68 (163)

FIGURE 8.13 Changing the map classes changes our understanding of how concentrated or dispersed the non-White British groups were in England according to the 2011 Census data

Colour palettes

The reason for choosing four map classes in Figure 8.13 is that the map is shaded sequentially from the lowest to the highest values and the difference between the shading becomes hard to distinguish once there are more than four or five categories, especially if the maps are printed out in black and white (grey scale). The maps are designed to highlight more of something; that is, a greater percentage of

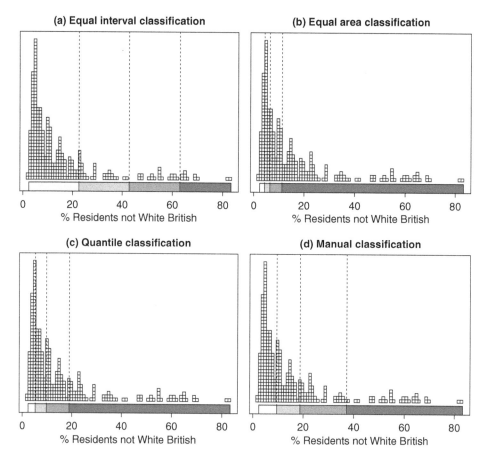

FIGURE 8.14 Showing the map classes used in Figure 8.13 and how they relate to the distribution of the variable percentage of residents not White British

the non-White British. The colour palette is one of the sequential schemes listed at http://colorbrewer2.org and is regarded as colour-blind safe.

Figure 8.15 adopts a different approach. Firstly, the local authorities have been ranked from the one with the least percentage of non-White British residents (rank 1) to the one with the greatest (rank 326). Second, the map has been shaded using a continuous colour ramp as opposed to discrete map classes. Finally, a diverging colour palette has been used, emphasising both the lowest and highest ranks.

A third type of colour palette is for 'qualitative' (i.e. categorical) data. This is used when there is no quantitative ordering to the data and the colour scheme should not imply that some values are numerically greater than others. The map at the beginning of this chapter (Figure 8.1) had a qualitative palette.

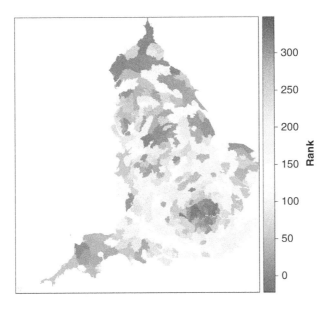

FIGURE 8.15 Showing the local authorities ranked by the percentage of their population that is not White British (highest rank = highest percentage). A colour ramp based on a diverging colour palette has been applied

Cartograms

It is time for a confession. The maps shown in Figures 8.13 and 8.15 are geographically dishonest in that they exaggerate the size of London and other urban areas. Taking Figure 8.13d as an example, it should really look like Figure 8.16a if the map is to accurately reflect the size (area) and the distances between places. However, that is not the only goal. The map is intended to communicate which places have the greatest percentages of a population that is not White British. The problem is that those places are also the smallest, despite the fact that a lot of people live within them. Their lack of geographical size makes them hard to see on the map.

A popular solution to this problem is the use of cartograms.[12] These resize the places in accordance with something other than their physical area. For example, Figure 8.16b has them sized according to the number of people who live within them. The result is to make London and other urban locations much more prominent. Another approach is to expand the areas according to the count of the variable of interest. In Figure 8.16c it is the number of people not White British. This is not simply a duplication of the shading: the colour indicates the *percentage* of the

---

[12]The ones shown here were produced using the Java version of ScapeToad: http://scapetoad. choros.ch

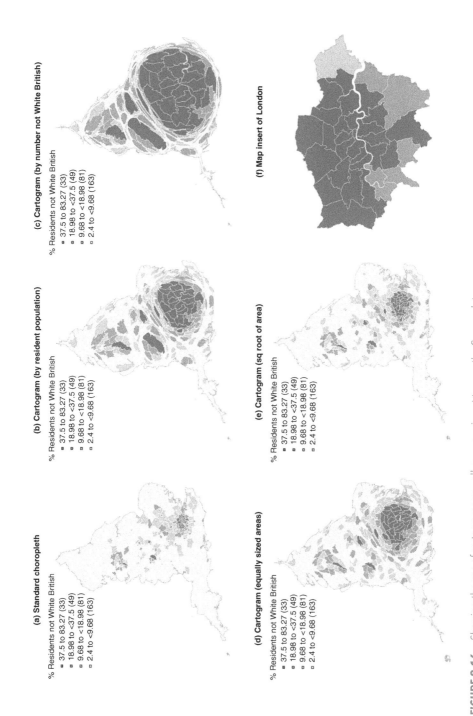

(a) Standard choropleth

% Residents not White British

- 37.5 to 83.27 (33)
- 18.98 to <37.5 (49)
- 9.68 to <18.98 (81)
- 2.4 to <9.68 (163)

(b) Cartogram (by resident population)

% Residents not White British

- 37.5 to 83.27 (33)
- 18.98 to <37.5 (49)
- 9.68 to <18.98 (81)
- 2.4 to <9.68 (163)

(c) Cartogram (by number not White British)

% Residents not White British

- 37.5 to 83.27 (33)
- 18.98 to <37.5 (49)
- 9.68 to <18.98 (81)
- 2.4 to <9.68 (163)

(d) Cartogram (equally sized areas)

% Residents not White British

- 37.5 to 83.27 (33)
- 18.98 to <37.5 (49)
- 9.68 to <18.98 (81)
- 2.4 to <9.68 (163)

(e) Cartogram (sq root of area)

% Residents not White British

- 37.5 to 83.27 (33)
- 18.98 to <37.5 (49)
- 9.68 to <18.98 (81)
- 2.4 to <9.68 (163)

(f) Map insert of London

FIGURE 8.16 Showing the use of cartograms as well as a map insert to map the Census data

population that is not White British; the size of the areas indicates the *number*. The resulting map gives even greater emphasis to where those groups are located.

Cartograms can be used to create unconventional, sometimes provocative and at times unsettling views of the world – see, for example, Benjamin Hennig's website at http://www.viewsoftheworld.net. They can be particularly effective when the aim is to draw attention to how unevenly something is distributed. For example, there is an interesting cartogram showing how many scientific research papers each country produced, numbers that are strongly biased towards the Northern hemisphere and against the global South, including Africa.[13] This says something of interest about the post-colonial nature of academic knowledge production.

Nevertheless, there is a problem: as the geography of the map becomes more contorted, it loses geographical detail and it becomes harder to work out exactly which place is which. Compromise solutions are shown in Figures 8.16d and 8.16e. The first gives each local authority equal space on the map – they are all approximately the same size (but not the same shape, although that is a possibility too[14]). The second resizes the places according to the square root of their actual areas. This was the one used for Figures 8.13 and 8.15. It has the effect of enlarging the smaller places relative to the largest, but with less geographical distortion. A more traditional solution is to use a map insert. In many respects, I prefer it.

8.9 CONCLUSION

There is much more that could be written about Geographical Information Systems and also about geographical information science, the second phrase pointing away from the software and towards the methods, principles and theories that underpin GIS-based analyses. There are plenty of textbooks on the subject, and a very useful resource is the online edition of *Geospatial Analysis*.[15] Some of the ideas will be explored further in Chapters 10 and 11.

Although they are geographical by name, GIS have not always been wholly welcomed, especially in human geography. An early criticism of GIS was that they are tools of the state, offering a very partial view of the world; that they act to objectify the world without giving voice or, rather, appearance to the plurality of often contested meanings that infuse the places around us. There is truth in this accusation. Nevertheless, although official maps of cities and states may be, for better or for worse, the products of government agencies, the same tools, and mapping more

---

[13] https://theconversation.com/its-time-to-redraw-the-worlds-very-unequal-knowledge-map-44206

[14] See, for example, https://richardbrath.wordpress.com/2015/10/15/equal-area-cartograms-and-multivariate-labels/

[15] http://www.spatialanalysisonline.com

generally, can be used to create counter-views and to assign different meanings to the spaces found on a map. This is the spirit of community mapping, participatory GIS and some aspects of counter-tourism.[16] The geographer, James Cheshire, and the designer, Oliver Uberti, have harnessed a wide range of data about London, providing what their book describes as '100 maps and graphics that will change how you view the city' (Cheshire and Uberti, 2012). Katherine Harmon's book, *You Are Here* (2003), focuses on maps that venture beyond the boundaries of geography or convention. Maps have the power to persuade but also to challenge. The featured graphics in the journal *Environment Planning A* are free to access from its website, and many are cartographic.[17] They include 'The human planet' (Hennig, 2013), 'The virtual Bible belt' (Zook and Graham, 2010), 'Digital divide: the geography of Internet access' (Graham et al., 2012), 'Tweets by different ethnic groups in Greater London' (Adnan and Longley, 2013), and 'Lives on the line: mapping life expectancy along the London Tube network' (Cheshire, 2012). They are well worth a look.

---

[16]See, for example, http://www.participatorymethods.org/method/participatory-geographical-information-systems-pgis, http://www.ppgis.net and the videos at https://vimeo.com/channels/pgis

[17]http://epn.sagepub.com/site/Virtual_Pages/Featured_Graphics_1.xhtml

9

LOOKING AT RELATIONSHIPS AND CREATING MODELS

9.1 INTRODUCTION

This and the following chapter are about creating models to explore the relationships between two or more data variables. There are many types of model that are used in geography. One is a literal model, something you could physically touch: a three-dimensional, scaled-down model of terrain, for example, or a wave flume in a laboratory to model the effects of waves on coastal erosion. Another is a digital model: a digital elevation model, for instance, or the reproduction of a cityscape using virtual reality technologies. A third is a computational model – a simulation – modelling flooding scenarios or processes of segregation and urban change. The Schelling models are a very well-known example,[1] and new methods of spatial microsimulation are extending our ability to create geographically detailed data sets that can be used to evaluate 'what if?' scenarios for public policy and commercial decision making.[2]

These chapters focus upon a fourth type: statistical models. They look at a set of measurements and ask if the variation found in one variable is related to the variation found in the others – whether one thing is linked to another. If it is, and if that relationship can be quantified accurately, then the model can provide the basis for prediction and potentially for explanation of what causes what.

This chapter gives attention to a bivariate model where there are only two variables to consider. One of these is the X variable, also known as the predictor

[1]These models, by Thomas Schelling (1969, 1971), showed how a slight preference for living with neighbours of the same ethnic group could lead to high levels of segregation.
[2]Lovelace and Dumont (2016).

or independent variable. The other is the Y variable, also called the response or dependent variable. Correlation analysis makes no meaningful distinction between the two, but regression analysis implies that a change in the X variable leads to a change in the Y. It may not be directly causative, and the association may arise through other mediating variables, but the direction of change is taken to be from the X to the Y, not the other way around. Whereas the correlation of X and Y is the same as the correlation of Y and X, a regression that models Y as a function of X is not the same as one that models X as a function of Y.

9.2 CORRELATION

Figure 7.14, in Chapter 7, used a scatter plot to show the relationship between domestic natural gas prices and electricity prices in selected European countries. A line of best fit was added to summarise the relationship. That line was found to be upward sloping because above average prices of gas tend to associate with above average prices of electricity, and below average prices of gas with below average prices of electricity. The relationship can be described as positive (in a mathematical sense) because it has an upward gradient – the variables go in the same direction.

Figure 9.1 gives a different example where they don't: the line is downward sloping and the relationship is negative. In this example, above average rates of unemployment are associated with below average neighbourhood house prices, and below average levels of unemployment are associated with above average

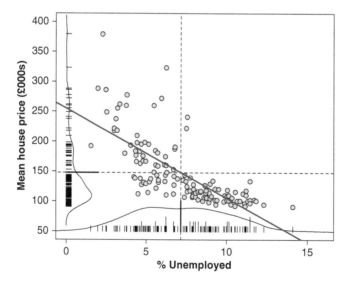

FIGURE 9.1 Showing the negative relationship between the neighbourhood average house price and the unemployment rate

house prices. This is not much of a revelation: employment and salary are key determinants of how much someone can afford to spend on a house. Affluence tends also to attract the sorts of cafés, shops, schools and other services that add value to a neighbourhood in a way that will be further capitalised in the house prices. The data, which are for Census zones in the city of Birmingham, England, are from the UK Land Registry Open Data service and from the 2011 Census.[3,4]

Although the neighbourhood average house price is related to the unemployment level, this is a vague observation that does not quantify the strength of the relationship. The two variables could be weakly or strongly related. Appealing to intuition, a strong relationship is where a small change in one variable has a big impact on the other, and where the relationship is unambiguous because the line fits the data well. In other words, a strong relationship is where the line of best fit has a steep gradient and where the observations do not vary greatly above or below it.

Figure 9.2 introduces a measure of correlation called the Pearson correlation.[5] Correlation is a statistical name for when two sets of data – two variables – appear related; when one seems to be dependent upon the other. The Pearson correlation statistic, usually denoted by r, ranges from −1, to +1. For these minimum and maximum cases, the line of best fit exactly fits the data and the gradient of the line is of no relevance beyond the fact that it slopes upwards for the positive correlation and downwards for the negative one: see the top row of Figure 9.2. The correlations are perfect because all the data points lie along the line.

As the data points scatter more widely around the line of best fit, so the magnitude of the correlation tends towards zero, which implies no correlation at all. This is shown in the middle row of Figure 9.2. A low correlation will arise when the line only weakly describes the relationship and the data points vary a lot above and below it.

Moving to the bottom row of Figure 9.2, the leftmost example illustrates a case of the line being almost flat and the correlation therefore close to zero. When the line is flat, it implies there is no relationship between the two variables because you cannot predict the Y value from the X value, or vice versa. They appear independent of each other. In fact, the gradient of the line is not quite zero in that example and so neither is the correlation. Indeed, it would be surprising with real-world data to encounter a circumstance when it was *exactly* zero. Note, however, how that the 95 per cent confidence interval ranges from −0.12 to +0.27. Chapter 5 discussed confidence intervals in the context of the standard error of the mean. Here they are based on the standard error of the correlation value, but the principle is the same: the standard error is a measure of uncertainty and from it the confidence interval

---

[3]http://landregistry.data.gov.uk/ and http://infuse.mimas.ac.uk/
[4]Specifically, the zones are middle layer super output areas, which have a mean average of 8129 persons living in them in Birmingham.
[5]Named after the mathematician and statistician, Karl Pearson.

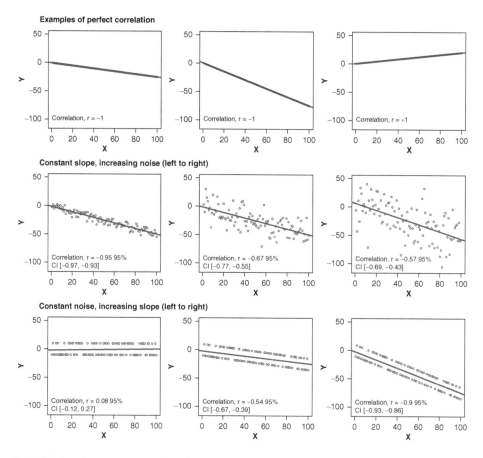

FIGURE 9.2 Showing various values of the Pearson correlation coefficient, *r*

is calculated. When the interval spans zero we cannot confidently dismiss the possibility that the 'true correlation' is zero and that the non-zero correlation we have observed is actually due to random variation (a sampling error) within the data. Here the confidence interval does span zero and the conclusion is that the correlation is not statistically significant at the 95 per cent confidence level. Even so, statistical significance (or the lack of it) ought to be of less interest than the correlation value itself: an *r* value of 0.08 seems pretty trivial, especially if we regard correlations of magnitude 0.1–0.3 as low, 0.3–0.5 as medium, and 0.5 or greater as high, as a rule of thumb. The correlation measures the effect size. At *r* = 0.08 it is very weak and in most contexts could probably be discounted.[6]

---

[6]But it does depend upon the context. Imagine a pill that offers only a slight improvement in life expectancy but is cheap to produce and has no harmful side effects. The pill is worth prescribing despite its weak effect.

Returning to the data in Figure 9.1, the correlation between the unemployment rate and the neighbourhood mean house price is $r = -0.723$ with a 95 per cent confidence interval ranging from -0.796 to -0.630. As discussed in Chapter 6, the meaning of this confidence interval is dubious if it is meant to convey that a correlation of the observed magnitude or more would be expected in less than five out of 100 random samples of the population when the true correlation is actually zero. It is dubious – some would say plain wrong – because the data are not random samples of a population. The Census is intended to be an enumeration of the entire population, and the house price data should include every sale made. However, that does make the data complete in the sense they are perfect and immune to error. It is moot whether the errors and uncertainties that are inevitable in the data approximate to randomness, but hopefully we are not using data that are biased in some systematic way as they would be if we excluded the sale of terraced properties or enumerated only female residents. Whatever is made of the confidence interval, it reminds us that there is uncertainty in all data and their analysis. In this example, and whether we look at the correlation value or its confidence interval, it is reassuring analytically to know that everything points to the same conclusion: there is a high, negative correlation between levels of unemployment and the average price paid for housing in the neighbourhoods of Birmingham. There is statistical evidence for the relationship, and it is a relationship that makes logical sense.

Covariance and calculating the Pearson correlation coefficient

The Pearson correlation is so widely used that you will find it in almost any data analysis software. Its calculation is based on the covariance of two variables. Recall, from Chapter 5, that the variance (s^2) is a measure of how much one variable varies around its mean. It is calculated as

$$s^2 = \frac{\sum_{i=1}^{n}\left(x_i - \bar{x}\right)\left(x_i - \bar{x}\right)}{n-1} \tag{9.1}$$

The covariance is similar, except that it multiplies the variation around the mean of one variable by the variation around the mean of the other:

$$\mathrm{cov}\left(X,Y\right) = \frac{\sum_{i=1}^{n}\left(x_i - \bar{x}\right)\left(y_i - \bar{y}\right)}{n-1} \tag{9.2}$$

The covariance will generate a high, positive number when high values in the X variable pair with high values in the Y, and when low values in the X pair with low values in the Y (a positive relationship). A high, negative covariance will occur when high–low and low–high values pair together (a negative relationship).

The covariance of the house price and unemployment variables is −104.8 when the unemployment rate is measured as a percentage and average house price in thousands of pounds. It decreases to −1.048 if the unemployment variable is converted to a proportion, and rises to −104,784 if the house prices are in pounds. It is not ideal to have a measure that changes depending upon the scale of the measurement units used. The correlation coefficient acts to standardise the covariance into the range −1 to +1, and does so by dividing the covariance of X and Y by the standard deviation of X multiplied by the standard deviation of Y:

$$r_{xy} = \frac{\text{cov}(X,Y)}{1/(n-1)\sqrt{\sum_{i=1}^{n}(x_i - \bar{x})^2}\sqrt{\sum_{i=1}^{n}(y_i - \bar{y})^2}} \qquad (9.3)$$

which simplifies[7] to

$$r_{xy} = \frac{\sum_{i=1}^{n}(x_i - \bar{x})(y_i - \bar{y})}{\sqrt{\sum_{i=1}^{n}(x_i - \bar{x})^2}\sqrt{\sum_{i=1}^{n}(y_i - \bar{y})^2}} \qquad (9.4)$$

and also to

$$r_{xy} = \frac{\sum_{i=1}^{n} z_x z_y}{n-1} \qquad (9.5)$$

where z_x and z_y are z-values: $z_x = (x - \bar{x})/s_x$, $z_y = (y - \bar{y})/s_y$ (see Section 5.5).

9.3 ISSUES AFFECTING THE CORRELATION COEFFICIENT

Non-linear relationships

The Pearson coefficient is a measure of linear correlation that presumes the variables have a straight-line relationship. That need not be the case. Figure 9.3 gives examples of when it isn't. In each there is a strong relationship between the two variables, it just isn't a linear one. Three of the examples are curvilinear; the fourth has three distinct groupings of data. The second and fourth correlations are particularly misleading. The U-shaped relationship produces an almost zero correlation, which

---

[7]Because the $n-1$ in the covariance and in the standard deviations cancel each other out.

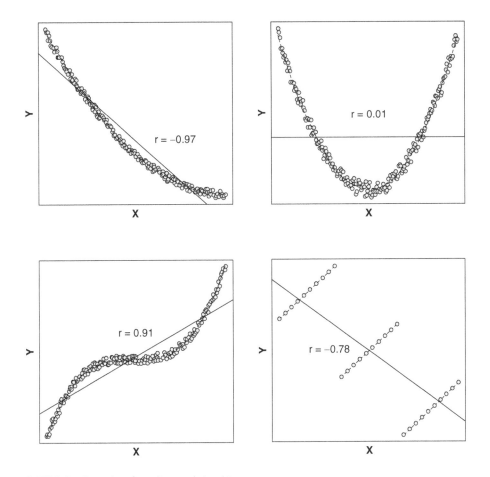

FIGURE 9.3   Examples of non-linear relationships

could wrongly be interpreted as saying there is no relationship. The fourth is an example of Simpson's paradox:[8] the three groups of data each have a positive correlation between the values of X and Y, but taken together they produce the reverse finding: a negative correlation. The moral of the story is to check the data with a scatter plot before applying the Pearson correlation coefficient to them.

A real-world example of Simpson's paradox was reported in the *New Statesman* following a post on the *New York Times* Economics blog (Hern, 2013). It considered how the median weekly wage in the USA, adjusted for inflation, had *risen* by 0.9 per cent from 2000 to 2013. However, the median wage had actually *fallen* for every subgroup of the population, regardless of whether it was high-school dropouts, graduates, those with some college education or those with a degree. The overall

---

[8]Named after a paper by the statistician, Edward Simpson (1951).

wage had increased, but for every group the average wage had decreased. How could this happen? Because there are now more college graduates and they, on average, get paid more than people in other jobs. More people in higher-paying jobs raises the overall wage but does not prevent those graduates from being paid less than in the past.[9]

Handling non-linear relationships

Looking back at Figure 9.1, and especially the left-hand side of the plot, we may suspect that the relationship between unemployment rates and the neighbourhood average house price is not linear: as unemployment decreases, the house price appears to rise faster and faster. The distribution of the Y variable has a long tail of higher values. Figure 9.4 fits a trend line with no assumption it will be straight. Instead it is allowed to bend to fit the data better. The details of the method are not important. What matters is that the result suggests the relationship is not linear.

There are a number of possible solutions to this problem, one of which is to see if a mathematical transformation will straighten the relationship. Figure 9.5 tries replacing the Y values with the logarithm of those values (yielding $r = -0.783$), and also replacing both the X and Y values with their logarithms ($r = -0.798$). Taking the square root of Y is not shown but is another possibility ($r = -0.756$).

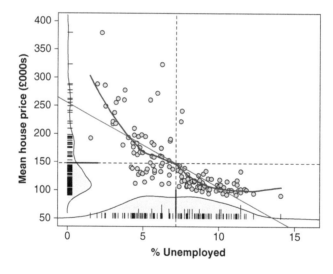

FIGURE 9.4 The relationship between the average neighbourhood house price and the unemployment rate appears not to be linear

---

[9]With a greater supply of people, perhaps companies don't need to pay so much to attract graduates to their jobs or perhaps jobs not formerly considered graduate level have become so.

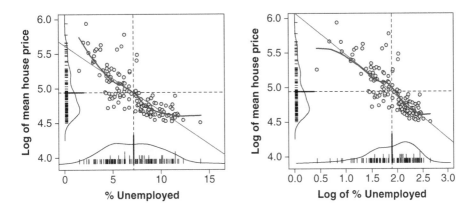

FIGURE 9.5 Transforming the data can work to straighten the relationship

A second approach is to see if the ranks of the values on the X-axis are correlated with the ranks of the values on the Y-axis; for example, do low-ranked values occur together, and do highly ranked values coincide also?[10] The advantage of this approach is that it ignores and therefore is not affected by the values becoming increasingly spread out as we go up along the Y-axis. Consequently, the measure of correlation, called Spearman's rank correlation coefficient (r_s), is less sensitive than the Pearson correlation to outliers in the tails of the data. For the Birmingham data, $r_s = -0.827$. As it happens, this is the strongest of all the correlations, but the goal is not to maximise the correlation: it is to use the right tool for the job, whether that be a transformation of the data or using a correlation other than the Pearson one.

The ecological fallacy, the modifiable areal unit problem and scale effects

Both the unemployment rate and the average house price variable are examples of neighbourhood data, where each value (each observation) describes a Census zone. Relationships observed at one scale of analysis need not apply at any other scale of analysis, and neighbourhood-level correlations do not necessarily reflect individual circumstances. It is possible that higher unemployment rates correlate with lower average house prices but the unemployed are themselves living in expensive housing. It is unlikely given the links between employment, salary and the value of a mortgage; nevertheless, the neighbourhood-level correlation does not, in itself, discount the possibility.

---

[10]It does not matter if the data are ranked from low to high or from high to low, provided it is done consistently for both variables. There are a number of ways to handle tied data (values in a variable that are the same and therefore have identical ranks). One is to assign them the average rank. For example, if ranks 4 and 5 are tied, rank them both as 4.5.

The ecological fallacy is to assume that a relationship found at a neighbourhood scale must necessarily apply to individuals living within those neighbourhoods. The seminal paper about this is William Robinson's (1950) analysis of 1930 US Census data, which showed a negative correlation between the proportion of overseas immigrants in a state population and the state illiteracy rate. Given there was an association between more immigrants and less illiteracy, the implication is that immigrants are more likely to be literate. However, the reasoning is fallacious. Robinson also showed that there was a positive correlation between an individual being an immigrant and the probability they were illiterate. In other words, immigrants were more, not less, likely to be illiterate. These seemingly contradictory findings, highlighted by the reversal of the neighbourhood-level correlation at an individual level, are explained by knowing that immigrants tended to live in places where non-immigrants had higher levels of literacy. The immigrants were not themselves the people with higher literacy levels but they were living alongside those who were.

The ecological fallacy is a logical fallacy – the assumption that something that has been measured (in this case, correlation at a neighbourhood scale) must necessarily apply to something that has not (correlation at an individual scale). In an ideal world we would always have the data to check that the individual-level relationships hold true. However, there are often good reasons why individual data are not available. Protecting anonymity and maintaining personal data protection are amongst them. Even if we are careful to avoid the ecological fallacy, it still needs to be remembered that statistical correlations, as well as other analyses of geographical data, are affected by the scale of the analysis: the number and geographical size of the units into which the data are aggregated. To demonstrate this, Table 9.1 shows how the correlation between the unemployment rate (X) and the percentage of households without a van or car (Y) varies from $r = 0.840$ (and $r_S = 0.875$) at the coarse scale of the 35 English counties, to $r = 0.635$ (and $r_S = 0.638$) at the small area scale of the 181,408 Census output areas. Generally, the greater the level of data aggregation, the greater the magnitude of the correlation. That is because aggregating the data has the effect of smoothing out the detail. Geographical differences that are present at smaller area scales are lost as the data are grouped into larger and fewer areas. This is evident by looking at the variances of each variable. The data are noisier at the small area scale than they are at the county scale.

Nevertheless, it cannot be assumed that the correlation will always be strongest at the coarsest scale of analysis. Ultimately the correlation depends upon the design of the analytical units and how well their geography replicates the real-world patterning of whatever is being studied. The modifiable area unit problem (MAUP) says that the scale of the analytical units could (in principle) be changed and, if it were, we can expect the results to change as well. Furthermore, even if the scale were fixed, the boundaries of the units could still be altered whilst keeping their overall size the

TABLE 9.1 Showing how the correlation between unemployment rate (X) and the percentage of households with neither a car nor a van (Y) changes with the scale of analysis

Census geography (level of aggregation)	n	Pearson correlation	Spearman's rank correlation	Variance of X	Variance of Y
Counties	35	0.840	0.875	0.962	75.9
Local authorities	346	0.760	0.858	1.50	105
Middle super output areas	7,201	0.768	0.828	4.26	219
Lower super output areas	34,753	0.743	0.777	5.89	267
Output areas (Census small areas)	181,408	0.635	0.638	9.52	342

same. That would change the results too. Inevitably, any analytical result is dependent upon the way the data are measured and collected. This would matter less if populations and other spatial distributions divided neatly into clearly defined geographical zones for which data could then be collected. Unfortunately, they rarely do, which means the geographies for which data are available are somewhat arbitrary whilst at the same time affecting the conclusions drawn from the data.

9.4 REGRESSION

Principles of regression

Correlation is used to determine whether one variable is dependent upon another, giving a measure of the strength of the relationship. However, it cannot be used to predict the value of Y for a given value of X. Ignoring the potential non-linearity for the moment, we know that the correlation between the two variables in Figure 9.1 is $r = -0.723$, 95% CI [−0.796, −0.630]. It's a strong relationship, but we cannot say from it what we should expect the mean house price to be if, for example, the unemployment rate is 10 per cent.

The line of best fit provides the information required to make the prediction. The equation of a line was considered in Chapter 4. The line shown in Figure 9.1 has the equation

$$\hat{y} = 254.8 - 14.96x \qquad (9.6)$$

where $\hat{y}$ ('y-hat') is the predicted or expected value for a given value of X. For an unemployment rate of 10 per cent, $x = 10$ and so $\hat{y} = 254.8 - (14.96 \times 10)$, which is 105.2 thousand pounds (i.e. £105,200). This could also be written as $E(y \mid x = 10) = 105.2$, which translates as the expected value of Y given that $x = 10$. More generally, the equation of the line can be written as

$$\hat{y} = \beta_0 + \beta_1 x \qquad\qquad (9.7)$$

The beta values are the parameters that define the line: β_0 is the Y-intercept (the value of Y when X is zero); β_1 is the gradient. That gradient is negative in equation (9.6) because the variables are negatively correlated. The task of regression analysis is to estimate these parameters.[11] Because the estimates are dependent upon the 'sample' of data from which they are estimated, we denote them as $\hat{\beta}_0$ and $\hat{\beta}_1$, respectively (betas but with added 'hats').

Before it can be estimated, we need to define what we mean by a line of best fit. There are various possibilities but, intuitively, it is the one that passes through the centre of the cloud of data points shown in the scatter plot, and which has the least possible variation around it. For every observation in the data we obtain three values: the observed value of X, the observed value of Y, and the expected value of Y given the value of X. The observed values are those that are in the data (the actual measurements). The expected value is the value of Y predicted by the equation of the line. The line of best fit can be defined as the one for which the predicted values of Y come as close as possible to the actual values of Y; where what is expected of Y best meets what is observed.

Specifically, in what is called ordinary least squares (OLS) regression, the line of best fit is the one that minimises the square of the difference between the observed (Y) and the expected values ($\hat{y}$),

$$\min \sum_{i=1}^{n} \left(y_i - \hat{y}_i \right)^2 \qquad\qquad (9.8)$$

which, because $\hat{y}_i = \hat{\beta}_0 + \hat{\beta}_1 x_i$ (equation 9.7), can also be written as

$$\min \sum_{i=1}^{n} \left(y_i - \hat{\beta}_0 - \hat{\beta}_1 x_i \right)^2 \qquad\qquad (9.9)$$

Having defined the line of best fit in this way, the mathematical method of differentiation can be applied to equation (9.9) to generate two further equations, one for each of the parameters.[12] The first is

$$\hat{\beta}_0 = \bar{y} - \hat{\beta}_1 \bar{x} \qquad\qquad (9.10)$$

---

[11] It is called 'regression' because of how the children of very tall or short parents have heights that regress towards the mean height of the population (see Chapter 2).

[12] It would be a long digression to derive the regression equations here, but the various steps can be found online or in many statistical textbooks. The basic process is to differentiate equation (9.9) with respect to $\hat{\beta}_0$ and to $\hat{\beta}_1$, set the results equal to zero (because we are determining a minimum) and then rearrange the equations.

which indicates that the regression line always goes through the point $(\bar{x}, \bar{y})$, the mean of X and Y. The second is

$$\hat{\beta}_1 = \frac{\sum_{i=1}^{n}(x_i - \bar{x})(y_i - \bar{y})}{\sum_{i=1}^{n}(x - \bar{x})^2} \qquad (9.11)$$

which is equal to

$$\hat{\beta}_1 = \frac{\left(\sum_{i=1}^{n}(x_i - \bar{x})(y_i - \bar{y})\right)/(n-1)}{\left(\sum_{i=1}^{n}(x - \bar{x})^2\right)/(n-1)} \qquad (9.12)$$

and so

$$\hat{\beta}_1 = \frac{\text{cov}(X, Y)}{s_x^2} \qquad (9.13)$$

The gradient is the covariance of X and Y divided by the variance of X.

Regression in matrix form

The equation of the regression line, equation (9.7), can be written more compactly as

$$\hat{\mathbf{y}} = \mathbf{X}\hat{\beta} \qquad (9.14)$$

where $\hat{\mathbf{y}}$ is a column vector of predicted Y values, $\mathbf{X}$ is a matrix that contains the X values, and $\hat{\beta}$ is the beta values (the Y-intercept and gradient). Continuing in matrix form, these values are calculated as

$$\hat{\beta} = \left(\mathbf{X}^T\mathbf{X}\right)^{-1}\mathbf{X}^T\mathbf{y} \qquad (9.15)$$

For the data in Figure 9.1,

$$\hat{\mathbf{y}} = \begin{bmatrix} & (\pounds000s) \\ (\hat{y}_1) & 207 \\ (\hat{y}_2) & 198 \\ (\hat{y}_3) & 217 \\ \cdots & \cdots \\ (\hat{y}_n) & 134 \end{bmatrix} \quad \mathbf{X} = \begin{bmatrix} & & (\%) \\ (x_1) & 1 & 3.19 \\ (x_2) & 1 & 3.78 \\ (x_3) & 1 & 2.53 \\ \cdots & \cdots & \cdots \\ (x_n) & 1 & 8.07 \end{bmatrix} \quad \hat{\beta} = \begin{bmatrix} (\beta_0) & 255 \\ (\beta_1) & -15.0 \end{bmatrix} \qquad (9.16)$$

From this, and using the matrix multiplication introduced in Chapter 4, the value predicted by the regression line for the first value in the data set is $\hat{y}_1 = (1 \times \beta_0) + (x_1 \times \beta_1) = (255) + (3.19 \times -15.0) = 207$. For the second, it is, $\hat{y}_2 = (1 \times \beta_0) + (x_2 \times \beta_1) = (255) + (3.78 \times -15.0) = 198$, and so forth down the columns. If it is not clear why the column of 1s appears in the $\mathbf{X}$ matrix, it may be helpful to note that $\hat{y}_1 = 1 \times \beta_0 + \beta_1 x_1$, $\hat{y}_2 = 1 \times \beta_0 + \beta_1 x_2$, $\hat{y}_3 = 1 \times \beta_0 + \beta_1 x_3$, etc.

Don't be put off by the matrix algebra. Instead, focus on what is obvious from Figure 9.1: the regression line is one of best fit, not perfect fit. The variations above and below the line are the differences between the observed and the expected values, referred to as the residual errors. They are residual in the sense that they are left over. They are the part of Y that is unexplained by its relationship with X.

Denoting these errors as $\hat{\varepsilon}$, then

$$\hat{\varepsilon} = \mathbf{y} - \hat{\mathbf{y}} \qquad (9.17)$$

the differences between the observed and predicted values; or, equivalently, by substation from equation (9.14),

$$\hat{\varepsilon} = \mathbf{y} - \mathbf{X}\hat{\beta} \qquad (9.18)$$

The task of OLS regression is to minimise $\varepsilon^T \varepsilon$, where $\varepsilon^T \varepsilon$ is known as the residual sum of squares (RSS) and is just another way of writing $\sum_{i=1}^{n} (y_i - \hat{y}_i)^2$. Rearranging the equation gives

$$\mathbf{y} = \mathbf{X}\hat{\beta} + \hat{\varepsilon} \qquad (9.19)$$

meaning the actual values of Y are those predicted by the regression line ($\mathbf{X}\hat{\beta}$) plus or minus the errors ($\hat{\varepsilon}$). It is 'plus or minus' because some of the values lie above the regression line and, for these, $\mathbf{y} - \hat{\mathbf{y}}$ is a positive number. For those that are below the regression line, $\mathbf{y} - \hat{\mathbf{y}}$ is a negative number. Positive residuals are those for which the regression line underpredicts the measured value. Negative residuals are those for which the regression line overpredicts. This is shown in Figure 9.6.

9.5 INTERPRETING REGRESSION OUTPUT

Table 9.2 shows the sort of output typical from statistical software for regression analysis. The estimates of β_0 and β_1 are obvious, and from them we obtain the regression line, $\hat{y} = 255 + 15.0x$. The standard error (se) is, as on other occasions, a measure of uncertainty in the (beta) estimate. It increases with the noisiness of the

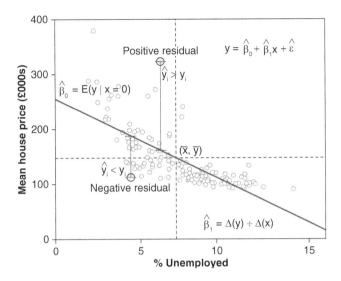

FIGURE 9.6 Showing positive and negative regression residuals

data (with the residual sum of squares, $\varepsilon^T\varepsilon$), and decreases with the number of observations, n. Dividing the beta estimate by the standard error produces the t-value, and the p-value is the probability of randomly drawing a t-value of greater magnitude than it from a t-distribution with $n - p$ degrees of freedom. In this second instance the p is not to be confused with the p-value but refers instead to the number of parameters estimated from the data. Currently there are two: the two beta values (the Y-intercept and the gradient).

For a more intuitive way to understand this, let's focus on the estimate of the gradient ($\hat{\beta}_1$). Recall from what we know about correlation that if the line of best fit is flat it implies no linear relationship between the Y and X variables. You will find an example of this if you look back at the bottom left chart of Figure 9.2. In this instance, knowledge of the X value provides no help in predicting the Y value because the values of Y bounce around independently of the X.

The key point is that zero gradient ($\hat{\beta}_1 = 0$) implies no (linear) relationship and that the X variable has no effect on the Y. The estimate of $\hat{\beta}_1$ is not zero in Table 9.2 but is −15.0. That means a one-unit increase in the X variable (the unemployment rate) is expected to lead to a 15-unit decrease in the Y variable (the neighbourhood average house price). But what if the 'true value' of the relationship was actually zero and the −15.0 has arisen due to sampling error? The p-value is the probability we have incorrectly dismissed a gradient of zero when it was actually true. In the table that probability is tiny and so it would be usual to reason that the relationship is statistically significant.

Before commenting on that reasoning, let's first extend it to the estimate of the Y-intercept ($\hat{\beta}_0$). Again, we are testing against a null and nil hypothesis: that the true Y-intercept is zero. This is unlikely given the p-value. Whether it matters or not depends upon the context. Since it is implausible that house prices would cost nothing in areas of no employment, it is reassuring to discover that the Y-intercept is 'significant'. However, imagine a model of the number of crimes committed against tourists in city neighbourhoods where the number of tourists visiting the neighbourhood is used as the predictor variable. In this example, there ought to be no crimes when there are no tourists, so we would not want the estimate of the Y-intercept to differ significantly from zero.

The standard errors reported in Table 9.2 can be used to create a confidence interval around the beta estimates. As a guide, $\hat{\beta} \pm 2 \times se_{\hat{\beta}}$ will be approximately equal to a 95 per cent confidence interval, provided there are more than about 50 observations in the data. For $\hat{\beta}_0$ this gives a confidence interval from about $255 - 2 \times 9.56$ to $255 + 2 \times 9.56$, which is from 236 to 274, and for $\hat{\beta}_1$ it is from about $-15.0 - 2 \times 1.25$ to $-15.0 + 2 \times 1.25$, which is from -17.5 to -12.5. Your statistical software will be able to calculate the 95 per cent or any other confidence interval more exactly this this. In the current example the confidence intervals do not include zero, meaning both beta values are significantly different from zero at a 95 per cent confidence level (a fact that can also be deduced from their p-values being less than 0.05: $p < 0.05$ means 95% confidence; $p < 0.01$ means 99% confidence).

Here, as on other occasions, we encounter the issue of what the confidence intervals (and also the p-values) mean for data that are not a random sample of a larger population. One answer is to say that the confidence intervals provide a means of expressing the uncertainty in the estimates of the beta values given the variability of the data. Another is to treat them with a degree of caution and without the traditional reverence that p-values in particular have been afforded in regression analysis. In fact, even if the data were a random sample, we know (from

TABLE 9.2 Typical regression output, here from a regression of neighbourhood average house price against the unemployment rate for neighbourhoods in Birmingham, England

	estimate	se	t-value	p-value
β_0	255	9.56	26.6	< 0.001
β_1	−15.0	1.25	−11.9	< 0.001
Residual standard error: 3.80 on 130 degrees of freedom				
R^2	0.522			
F-statistic: 142.3 on 1 and 130 degrees of freedom, p-value < 0.001				

Chapter 6) that a cynic can regard the p-values as little more than an elaborate measure of the number of observations in the data set: as n increases then, all other things being equal, the standard errors decrease, the t-values increase, the p-values drop and the confidence intervals narrow. The consequence is that, given a large enough data set, almost everything will be 'statistically significant'.

We address this by considering the effect size as well as how well the regression line fits the data. The effect size is the gradient of the line: by how much we expect Y to increase or decrease given a one-unit change in X. The estimate of β_1 in Table 9.2 says that for every percentage point increase in the unemployment rate, the average neighbourhood house price is expected to fall by about 15 thousand-pound units (i.e. £15,000). That equates to a predicted house price difference of £187,726 between the areas of lowest and highest unemployment, which is a lot. Now imagine that $\hat{\beta}_1$ had been 0.15 instead of 15. That would indicate a £150 decrease in the average house price for every percentage point increase in unemployment. Given enough data, that value could still test as statistically significant, but the effect of unemployment rates on the average house price now seems trivial: just because something is statistically significant does not mean it is substantively so (nor is a result that is statistically insignificant necessarily substantively unimportant). The point is that we cannot just rely on measures of statistical significance to decide whether the effect of one variable on another matters or not.

How well the regression line fits the data is measured by the R^2 value, also called the coefficient of determination. It gives the proportion of the variance in the Y value that is explained by the regression predictions:

$$R^2 = \frac{\sum_{i=1}^{n} \left(\hat{y}_i - \overline{y} \right)^2}{\sum_{i=1}^{n} \left(y_i - \overline{y} \right)^2} \tag{9.20}$$

As its name implies, it is the Pearson correlation of the X and Y values squared, $R^2 = r_{xy}^2$. It is also the correlation between the observed and expected Y values squared, $R^2 = r_{y\hat{y}}^2$. The R^2 value of 0.522 indicates that 52.2 per cent of the variance in the neighbourhood average house prices is explained by the unemployment variable, which is a sizeable amount.

Of the remaining statistics in Table 9.2, the residual standard error, $\hat{\sigma}_\varepsilon$ measures the spread of the residuals around the regression line. Figure 9.7a below shows the distribution of the residuals. To the left are the Y values that lie furthest below the regression line and are negative residuals. To the right are the values that lie furthest above the regression line and are positive residuals. Most lie somewhere in-between, occupying a position closer to the regression line and therefore closer to zero. The residual standard error is the standard deviation of this distribution:

$$\hat{\sigma}_\varepsilon = \sqrt{\frac{\Sigma_{i=1}^n \left(y_i - \hat{y}_i\right)^2}{n-p}} = \sqrt{\frac{\hat{\varepsilon}^T \hat{\varepsilon}}{n-p}} = \sqrt{\frac{\text{RSS}}{n-p}} \qquad (9.21)$$

where $n-p$ is the degrees of freedom.

Finally, the F-statistic derives from what is known as an analysis of variance (ANOVA) and is a test of whether the regression line leads to a reduction in the residual sum of squares when compared to a simpler null model. The idea of the null model is that the Y values could be predicted without reference to the X values, using just the mean of Y instead. That is, we could assume that $\hat{\mathbf{y}} = \bar{y}$ for all y instead of making the regression prediction $\hat{\mathbf{y}} = \mathbf{X}\hat{\boldsymbol{\beta}}$. The assumption would be wrong because none of the Y values is equal to its mean. However, the issue is whether the residual sum of squares from the null model, $\text{RSS}_0 = \left(\mathbf{y} - \bar{y}\right)^T \left(\mathbf{y} - \bar{y}\right)$ is much greater than the sum of squares from the regression model, $\text{RSS}_1 = \left(\mathbf{y} - \mathbf{X}\hat{\boldsymbol{\beta}}\right)^T \left(\mathbf{y} - \mathbf{X}\hat{\boldsymbol{\beta}}\right)$. If not, then the inclusion of the X variable in the model is not explaining much of the variation in the Y values, suggesting they are unrelated.

The F statistic is

$$F = \frac{\left(\text{RSS}_0 - \text{RSS}_1\right)/\left(p-1\right)}{\text{RSS}_1/\left(n-p\right)} \qquad (9.22)$$

where the associated p-value provides a benchmark for rejecting the null model in favour of the regression model. This may seem to be duplicating information that we can already discern by looking at the t- and p-values associated with the estimate of the slope, $\hat{\beta}_1$. Indeed, $F = t_{\hat{\beta}_1}^2$ and their p-values are the same. However, the statistics diverge when we introduce more than one X variable into the model in an effort to better explain Y. In that situation (considered in Chapter 10), whereas the t-statistics look at each slope separately and test if it might be zero, the F statistic looks at the slope values jointly and asks whether they *all* could be zero. In other words, the F statistic considers whether any aspect of the model says something useful about Y, whereas the t-values look at specific variables.

9.6 REGRESSION ASSUMPTIONS

For the statistical tests to be valid, assumptions are made about the residuals from OLS regression: namely that they are independent of each other and are normally distributed around the regression line with a mean of zero and a constant variance.[13] You may see this written as $N(0, \sigma^2)$ or in similar notation. It says that the expected difference between $\hat{y}$ and y should be zero, regardless of the value of X: $E(\hat{y} - y) = 0$ for all x. More simply, it means that the residuals should be random noise, exhibiting no structured pattern and unrelated to either the X or the predicted values, $\hat{y}$. The predicted values are also known as the fitted values.

Visual tools are helpful to check if these assumptions have been met. Figure 9.7a shows the distribution of the residuals as a histogram and compares it to a normal distribution. Figure 9.7b is a density plot that is very similar to the histogram but doesn't have its 'blockiness'. Figure 9.7c is a quantile plot that compares the actual values of the residuals with what they would equal if they were normally distributed. If the residual distribution were perfectly normal, all the points in the quantile plot would lie along the dotted line, although some deviation at the ends of the line is to be expected given that the normal curve extends from minus to positive infinity, which is unlikely for a set of measurements. In this case, all the plots are revealing the same thing: the residuals have a tail of high values – neighbourhoods where the houses are more expensive and the regression line is underpredicting their average house price. Finally, Figure 9.7d plots the residuals against the predicted (or fitted) values. Note how the residuals are varying more greatly around the zero line for the higher values to the right of the plot than they are for the lower values to the left. This is a situation known as heteroscedasticity and it violates the assumption of a constant variance.[14] The residuals should be identically and independently distributed (sometimes referred to as i.i.d.). They are not: their variance is not identical across all the fitted values and there is a pattern which violates the observation of independence: observe how the trend line turns upwards because the highest predicted values tend also to be underpredictions.

With the regression assumptions violated, the question is what to do about it. One possibility is to do nothing: it is the standard errors, confidence intervals and p-values that are affected. If we are not especially interested in the statistical significance of the model (or question the legitimacy of what that means when applied to our data) then maybe the violations can be ignored. In fact, even if we are interested, unless the violations are serious, the regression statistics are generally

---

[13]More correctly, the assumptions apply to the errors, ε, of the 'true model', $\mathbf{y} = \mathbf{X}\beta + \varepsilon$, not what is being estimated, which is $\mathbf{y} = \mathbf{X}\hat{\beta} + \hat{\varepsilon}$, the issue being that the regression residuals, $\hat{\varepsilon}$, are not the same as ε. This is a subtle distinction that relates to ideas of sampling and randomness in the data. Here it is sufficient to appreciate that whereas the $\hat{\varepsilon}$ are measured from the data, the ε are unknown. Testing the assumptions as they apply to the residuals is the closest we can get to testing them on the ε.

[14]My publisher prefers heteroscedastic (with a c) to heteroskedastic (with a k) but I am not convinced they are correct to do so: see Paloyo (2014).

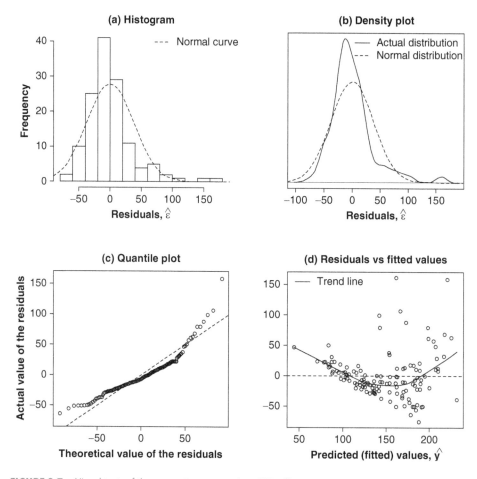

FIGURE 9.7 Visual tests of the regression assumption, N(0, σ^2)

robust to the assumptions not being full met. The normality assumption matters less as the data set increases in size, and technical fixes are available for the heteroscedasticity. However, the real issue is why the heteroscedasticity exists. What it points to is the violation of another, more fundamental assumption: the form of the model, estimating the mean neighbourhood house price as equal to $\hat{\beta}_0 + \hat{\beta}_1 \times$ the unemployment rate. We have reason to doubt this model because we suspect the variables are not linearly correlated. As such, the model seems incorrectly specified. One alternative would be to fit a model of the form,

$$\log(\hat{y}) = \hat{\beta}_0 + \hat{\beta}_1 x \tag{9.23}$$

by fitting the regression line to the log of the Y values. Another is

$$\hat{y} = \hat{\beta}_0 + \hat{\beta}_1 x + \hat{\beta}_2 x^2 \tag{9.24}$$

Although these models are both positing a non-linear relationship between X and Y, they still count as linear models and can be fitted with OLS regression. All we are doing is transforming the X or Y variable in some way: equation (9.23) takes the log of the Y variable (and could take the log of the X variable too); equation (9.24) fits a polynomial relationship by taking powers of the X variable. Of the two, an advantage of the second approach is that because $\hat{y} = \hat{\beta}_0 + \hat{\beta}_1 x$ is a sub-model of $\hat{y} = \hat{\beta}_0 + \hat{\beta}_1 x + \hat{\beta}_2 x^2$, we can use an analysis of variance with an F test to see if there are significant gains in including the additional variable, x^2. It turns out that there are, $F = 18.5, p < 0.001$. The estimated coefficient for the new predictor is significant, $\hat{\beta}_2 = 1.73$, 95% CI $[0.933, 2.52]$, and the R^2 value rises to 0.583. Even so, there is still a tail of positive residuals corresponding to the neighbourhoods for which the model is underestimating the mean house price.

Extreme residuals and leverage points

Figure 9.8 shows a regression line around which there are two obvious outliers, marked A and B. Both of these can be regarded as extreme residuals, ones that are a long way above or below the line. Aside from checking for them on a scatter plot, another way of identifying an extreme residual is to convert the raw residuals to a standardised scale by dividing by the standard deviation of their distribution:

$$\varepsilon^* \cong \hat{\varepsilon}/\hat{\sigma}_\varepsilon \tag{9.25}$$

where ε^* are the standardised residuals and $\hat{\sigma}_\varepsilon$ is the standard error of the residuals.[15] As a rule of thumb, standardised residuals of magnitude greater than 2 may be a problem, but should not automatically be assumed to be so. Of more concern are points that are a long way from the other data points and have the effect of moving the regression line towards them. Point A in Figure 9.8 is an extreme residual but its effect is minimal, shifting the line only slightly upwards. Point B has a greater distortive effect, rotating the line. Although it is no further below the regression line than A is above it, point B is also a long way from the mean of X and acts like a lever, causing the line to pivot around $(\bar{x}, \bar{y})$. It is for this reason that point B is described as a leverage point.

---

[15] As a formula for calculating the standardized residual, $\hat{\varepsilon}/\hat{\sigma}_\varepsilon$ is not exactly correct. However, it illustrates the idea: the raw residuals are scaled with respect to the standard deviation of their distribution.

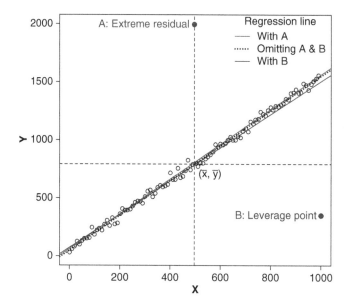

FIGURE 9.8 Illustrating two examples of an extreme residual. Only one is influential on the regression line and is a leverage point

By comparing the effects of A and B we learn that not all extreme residuals are necessarily influential upon the model fit. For those that are, there are three ways forward: check the model specification (fitting a straight line to what is actually an exponential relationship will generate extreme residuals, for example); omit the observations and refit the model; or use a method of regression that is robust to outliers (e.g. robust regression, resistant regression or quantile regression).

9.7 CONCLUSION

Although bivariate models have the virtue of simplicity, they can only take us so far. It would be surprising if the unemployment rate fully explained the average neighbourhood house price in Birmingham. Other correlates might include the type of properties found in the neighbourhoods, when they were sold, the distance from the city centre and the local crime rate, amongst other possibilities. Failure to include these variables could lead to a model misspecification, whereas including them should explain more of the variation in the neighbourhood prices.

A model that has more than one X variable to explain and predict the Y is known as a multiple regression model, and is looked at in the next chapter. That chapter also addresses another question, which is 'what is geographical about regression analysis?' To this point, the answer might be 'nothing!' Although we are using geographical (i.e. neighbourhood) data, nothing geographical about those data has featured in the analysis. Indeed, the assumption that the regression residuals are independently and identically distributed means that there should be no spatial patterning to them. That is possible, but it is also quite a bold assumption if the variable we are modelling has itself a strong spatial pattern – and, so far, we have not stopped to check. Therefore, as well as introducing multiple regression, the following chapter will introduce some more geographical methods of analysis that also set the scene for Chapter 11.

10

MULTIPLE REGRESSION AND GEOGRAPHY

10.1 INTRODUCTION

This chapter continues on from Chapter 9, extending the regression framework to include multiple predictor variables explaining the response variable. It discusses various model specifications, including for when the response variable is measured on a logarithmic scale or when it measures a binary (yes/no) outcome. It looks at the use of mean-centring and standardising the X variables to aid interpretation of the model.

Throughout the chapter, a focus is on taking a geographical approach to regression. Regression can be understood as trying to explain the patterns and differences we find in a map of geographical data. Although there is not necessarily a problem in using regression with geographical data, a complicating factor may arise when the geographical patterns we are trying to explain are very pronounced; when, for example, high or low values cluster together in different parts of the map, creating a link between where we look on the map and what we find there. To discover there is a geography to crime, unemployment, wealth, industrial location, soil types, animal species, flood risk, environmental pollution, land use, or whatever, is hardly surprising. The problem is when those geographies filter through into the model's residuals, violating the assumption that the residuals are independent of one another.

To some disciplines, spatial dependencies are a nuisance, undermining the validity of the model. But, for a geographer, detecting geographical patterns may be a first stage to understanding the processes that created them or to analysing the measurements within their spatial context. This chapter looks at ways of measuring what is known as spatial autocorrelation – the idea that geographically proximate values are correlated with one another. It uses a spatial weights matrix to define what is meant by neighbouring locations and, towards the end, adopts a more geographical approach to regression modelling that incorporates the spatial weights matrix within it.

It is quite a long chapter, and invites the question why so much space is devoted to the topic of regression analysis? The answer is that regression types of analysis are so widely used in science and social sciences that they deserve the coverage. Although this chapter only scratches the surface of multiple regression, it provides guidance on some of the core issues and ideas, and raises some geographical considerations that are considered further in Chapter 11.

10.2 MULTIPLE REGRESSION FROM A GEOGRAPHICAL PERSPECTIVE

Figure 10.1 offers a geographical perspective on what regression analysis aims to achieve, using the same data that we looked at in Chapter 9. The idea has been to explain the geographical patterns in the Y variable (the neighbourhood average house price) with reference to the geographical patterns in the X variable (the neighbourhood unemployment rate). However, the model was only partly success-ful in doing so. The map of the standardised residuals shows there are clusters of neighbourhoods where the average house prices are overpredicted (generating negative residuals) and, more especially, there are clusters where the average is underpredicted (positive residuals). Spatial patterns in the residuals violate the assumption that their values are independent of one another, and that violation affects the uncertainty around the estimated effect sizes. As a consequence, the standard errors of the estimated beta values typically are underestimated, which means the confidence intervals and the p-values are also underestimated (the t-values are overestimated).[1] If the p-value is underestimated, it becomes easier for an X variable to be judged statistically significant at a given threshold such as 95 per cent confidence, where $p < 0.05$.

Two important geographical considerations can be seen in the maps of Figure 10.1: spatial heterogeneity and spatial dependencies. Spatial heterogeneity is geographical variation across the study region. The neighbourhood average house price is not the same everywhere. Spatial dependencies are revealed by the patterns of clustering and sometimes expressed in Waldo Tobler's well-known 'first law of geography': 'Everything is related to everything else, but near things are more related than distant things' (Tobler, 1970). It's not really a law and is by no means universally true, but as a working assumption it is often validated in the real world – plant species cluster

---

[1]Recall that the standard error is a measure of uncertainty that decreases with an increase in the number of observations, n. If the values are not independent of each other then, in effect, we have less information then we thought and the standard error ought to be cal-culated based on a value less than n, which would make the standard error larger.

together, as do soil types, businesses, services, people with similar social or ethno-cultural backgrounds, crime, and so forth. It is a description of spatial autocorrelation, specifically positive spatial autocorrelation, when things that are alike are also closely located to each other. Negative spatial autocorrelation is when there is a pattern of opposites, as in the black-and-white pattern of a chessboard.

A pattern of spatial autocorrelation in the residuals is sometimes regarded as a 'technical problem' and a nuisance. However, it ought to be more than that, inviting the substantive question of how the patterning arose in the first place. It may point to a model misspecification. For example, Chapter 9 considered the curvilinear relationship between the neighbourhood average house price and the unemployment rate. A failure to allow for this may cause the pattern in the residuals. However, it also suggested that it was optimistic to hope that the variation in the Y variable could be explained by a single, predictor X variable. The model misspecification may be one of missing variables – other predictor variables that need to be brought into the picture to better explain the patterns of house prices. This is what multiple regression allows.

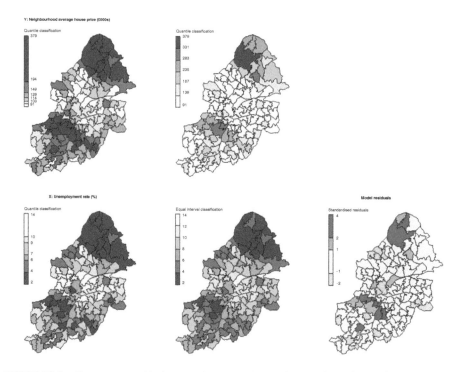

FIGURE 10.1   From a geographical perspective, regression analysis can be understood as trying to explain the map of the *Y* variable in relation to one or more *X* variables. Mapping the data also reveals evidence of spatial heterogeneity as well as spatial clustering, some of which remains after the model has been fitted

10.3 VARIABLE SELECTION

How do we select the variables to be included in the model? There are two approaches that could be adopted, one more theory-led and the other driven by the data. The first is characteristic of econometric analysis, using regression-based methods for testing economic models. Those models usually derive from economic theory, so the process of modelling starts with a theory, then a mathematical model of the theory, next an empirical/statistical model to approximate the mathematical model with the data available, and finally the results. It is a deductive process: the idea is to use the model to test (and deduce) whether the theory is correct. The theory comes before the data gathering.

The data–driven approach is associated with some areas of (geo)computation and applications of machine learning. It tries to make sense of the information contained in what is often a large data set. It is an inductive approach – from the data emerge new ideas and knowledge. When applied to regression analysis, throwing data at a model and seeing what sticks is generally frowned upon: given enough variables, it is more or less inevitable that some will appear statistically significant and will explain much of the variation in the variable of interest. That does not make the model meaningful nor the results substantively important. Traditionally, statisticians have valued parsimony, preferring simpler models with few variables over complex models with lots of them. Although there is clearly the possibility of oversimplification, a model ought to be treated with caution if it makes no obvious sense, cannot be explained or encourages *post hoc* rationalisations simply to fit the findings ('explanation on the hoof').

Many applications will sit somewhere between the two extremes, with a rationale for the choice of variable selection and a rough idea in mind but also an element of trial and error to see what works. For example, the population flow or interaction between two places, f_{ab}, can be theorised as a gravity model, $f_{ab} = m_a m_b / d_{ab}^k$, where m_a is the mass (importance/attraction) of place a, m_b is the same for place b, and d_{ab}^k is the distance between them raised to some power, k. The model says that the interaction between two places increases with their mass and decreases with their distance apart. Recalling the discussion about logarithms in Section 4.6, the model can be estimated as

$$\log\left(f_{ab}\right) = \beta_0 + \beta_1 \log\left(m_a\right) + \beta_2 \log(m_b) - \hat{\beta}_3 \log\left(d_{ab}\right) \qquad (10.1)$$

where $\hat{k} = \hat{\beta}_3$ and the additional parameters, β_0, β_1 and β_2 are assumed to equal 0, 1 and 1, respectively. Whether they do equal those values can be explored by fitting the model and looking at the estimated betas; alternatively, the model can be set up with those parameters constrained to equal their theoretical values but estimating the value of $\hat{k}$.[2] There is other room for exploration. For example, the analyst may wish to consider different ways of measuring how attractive a place is – using population size, economic productivity, or the range of goods and services, for example – and the distance apart could be a straight-line distance, a network distance along a transportation route, a measure of accessibility, a measure of ine- quality or a measure of how socio-culturally different the places are. The analyst may not be clear at the outset which variables will work best but will be trying out ideas and learning from the results in order to help form a theory.

10.4 AN EXAMPLE

Table 10.1 reports the results of a regression where eight additional X variables are used together with the unemployment rate to predict the neighbourhood average house price, Y. The logic is that price is a function of not only the residents' employment levels but also the characteristics of the properties, when they were sold, their location relative to the city centre, and the crime rate per thousand population. The assumption is that a greater percentage of detached properties in a neighbourhood will lead to a higher average because those properties occupy more land, new homes will attract a premium, that people value accessibility to the city centre positively, that house prices have risen over the period of the data (2012–14) and that higher crime rates suggest a less attractive and therefore cheaper neighbourhood. These ideas are expressed in a model,

$$\hat{y} = \beta_0 + \beta_1 x_1 + \beta_2 x_2 + \cdots + \beta_9 x_9 \tag{10.2}$$

where x_1 is the unemployment rate, x_2 is the percentage of properties sold that were detached properties, x_3 is the percentage that were flats or apartments, x_4 is the percentage of terraced properties, x_5 is the percentage that were new builds, x_6 is the percentage of the sales in 2013, x_7 is the percentage in 2014, x_8 is the distance from the city centre (in kilometres), and x_9 is the crime rate (per thousand popu- lation). In matrix form, this model is still expressed as $\hat{y} = \mathbf{X}\hat{\beta}$, and the beta values are still estimated as $\hat{\beta} = (\mathbf{X}^T\mathbf{X})^{-1}\mathbf{X}^T\mathbf{y}$, as in Chapter 9. The difference is that the $\mathbf{X}$ matrix now contains ten columns of data: one each for the nine predictor var- iables plus a column of ones to allow for the Y-intercept.

---

[2]To do so we can fit a regression line without an intercept term ($\beta_0 = 0$) and using the variables $\log(m_a)$ and $\log(m_b)$ as what are known as offsets (they are subtracted from the Y variable and therefore offset it by that amount).

TABLE 10.1 Regression output for a multiple regression model of neighbourhood average house prices

		estimate	se	t-value	p-value	95% CI
β_0	(Intercept)	129	44.4	2.91	0.004	[41.2, 217]
β_1	% Unemp	−10.2	1.21	−8.41	< 0.001	[−12.6, −7.81]
β_2	% Detached	3.12	0.272	11.5	< 0.001	[2.58, 3.66]
β_3	% Flat or apartment	0.790	0.189	4.19	< 0.001	[0.417, 1.16]
β_4	% Terraced	0.458	0.166	2.77	0.007	[0.130, 0.787]
β_5	% New build	0.134	0.225	0.595	0.553	[−0.312, 0.580]
β_6	% Sold 2013	0.566	0.622	0.909	0.365	[−0.666, 1.80]
β_7	% Sold 2014	0.684	0.524	1.31	0.195	[−0.354, 1.72]
β_8	Crime rate	−0.851	0.285	−2.99	0.003	[−1.41, −0.288]
β_9	Distance to city	−3.30	1.09	−3.03	0.003	[−5.45, −1.14]
		Residual standard error: 24.9 on 122 degrees of freedom				
	R^2	0.808		Adjusted R^2	0.794	
		F-statistic: 57.0 on 9 and 122 degrees of freedom, p-value: < 0.001				

Interpreting the output

Looking at Table 10.1, we find, as previously, that higher rates of unemployment are associated with lower average house prices. Continuing down the columns: an increase in the percentage of detached properties sold is associated with an increase in the average selling price, as is an increase in the percentage of flats or apartments, and an increase in the percentage of terraced properties. The greater the crime rate, the lower the average housing cost, and the further from the city centre, the lower the cost. The other variables are not statistically significant at the 95 per cent confidence level, although they still fit expectations: a greater percentage of new build properties raises the average selling price; sales from 2013 appear cheaper than sales from 2014. (Property prices tend to rise faster than inflation in the UK.)

The estimates of the beta values measure the expected change in the Y value given a one-unit change in the X variable, holding all the other X variables constant. It is the net effect of the X variable on the Y, once the effects of the other X variables have been taken out. For example, $\hat{\beta}_2 = 3.12$ means that for every one-unit increase in the percentage of houses sold that were detached, the neighbourhood average property price is expected to increase by 3.12 thousand-pound units (i.e. by £3120), assuming there is no change in the other variables. If that estimate seems overly exact, then we could turn to the 95 per cent confidence interval which predicts that the neighbourhood average will increase by between £2580 to £3660 for that one-percentage-point increase in detached properties sold. It is not routine for regression software to report the confidence interval, although it usually has the functionality to do so. Instead, it tends to give asterisks flagging how statistically significant or otherwise the reported beta values are. It has been argued that this makes for lazy analysis and lazy thinking – encouraging people to judge the success of the model solely by whether the X variables appear statistically significant or not and without a fuller consideration of the effect sizes and the overall model fit. Here the asterisks are omitted.

The presence of detached properties ($\hat{\beta}_2 = 3.12$) has less effect on the neighbourhood average than the unemployment rate ($\hat{\beta}_1 = -10.2$): a one-percentage-point increase in the unemployment rate is associated with a £10,200 decrease in house prices. Be careful when comparing effect sizes across variables: crimes per person and distance to the city centre are measured on a different scale than the other variables. Even those that are measured as percentages are distributed differently, with different means, ranges and standard deviations. They are not necessarily as comparable as they may first appear.

As with the simpler, bivariate regression in Chapter 10, the F-statistic shows that the model is an improvement over a null model with no predictor X variables. A separate analysis of variance (not shown) confirms it improves upon the model with just the unemployment variable in it.

Looking at the R^2 value, the current model explains 80.8 per cent of the variation in the Y variable, up notably from 52.2 per cent before (Table 9.2). Some increase is not surprising, it is inevitable – adding more X variables can only work to explain more of the variation in the Y variable. Because of this, because the increase can be driven by chance associations, and because a simpler model is preferred over a more complex one wherever possible, there is a second measure, called the adjusted R^2 value, that will only increase if the additional variable(s) really do enhance the predictive value:

$$R^2_{\text{adj}} = 1 - \frac{n-1}{n-p}\left(1 - R^2\right) \tag{10.3}$$

In the equation, the division by $n - p$ acts as penalty term: the greater the number of parameters, p (the number of beta values), the lower the adjusted R^2 value will be. Other measures of fit adopt the same idea. The Akaike information criterion (AIC) and the Bayes information criterion (BIC) penalise by $2p$ and by $p\log(n)$, respectively (see Faraway, 2015: Chapter 10). For the R^2 values, higher values indicate a better fit to the data. For the AIC and BIC, lower values do.

These measures of fit are shown in Table 10.2. They each point in the same direction: the residual error decreases for the multiple regression model, hence the R^2 values increase, and the two information criteria decrease. Each measure is saying

TABLE 10.2 Comparing measures of model fit for the simple bivariate and multiple regression models of neighbourhood average house prices

	Simple bivariate model	Multiple regression model	Change
Residual sum of squares, RSS ($\varepsilon^T \varepsilon$)	187,565	75,535	−112,030
R^2	0.523	0.808	+0.285
Adjusted R^2	0.519	0.794	+0.275
AIC	1339	1235	−104
BIC	1347	1267	−80

that the multiple regression model is better at explaining the neighbourhood average house price than the simple bivariate model. The added complexity of including the additional X variables appears justified, although we might wish to drop those that are statistically insignificant and/or have a small effect on the Y variable.

Regression diagnostics

Figures 10.2 and 10.3 use some of the visual methods discussed in Chapter 9 to check whether the regression residuals are independent and identically distributed. Figure 10.2 is for the initial, bivariate model. Figure 10.3 is for the multiple regression model. Comparing the top rows of the two, the additional X variables have reduced the tail of large, positive residuals, moving it closer towards a normal distribution. The heteroscedasticity also appears curtailed. A measure of heteroscedasticity is the Breusch–Pagan test, and it suggests the heteroscedasticity has been reduced but not eliminated.[3]

The bottom rows of the charts map the standardised residuals to look for evidence of clustering and therefore spatial dependence that would violate the assumption of independence. It is hard to assess by eye whether one map exhibits more clustering of the residuals than the other, which is why the Moran plot is useful. For each location on the map we have a standardised residual value, and we can also calculate the average residual for its neighbours. Calculating for each location in turn, we then have a set of paired values $(\hat{\varepsilon}_{(u,v)}, \overline{\varepsilon}_{(u,v)})$, where $\hat{\varepsilon}_{(u,v)}$ is the residual value at a location on the map denoted by (u, v) (which could be a grid

---

[3]The basis of the test is as follows. First, the residuals are obtained from the regression model and squared. Second, those squared residuals are taken as a Y variable and regressed against the same X variables that went into the model. The logic is that it should not be possible to predict the (square of) the residuals from the X_s if the residuals are unrelated to the X variables.

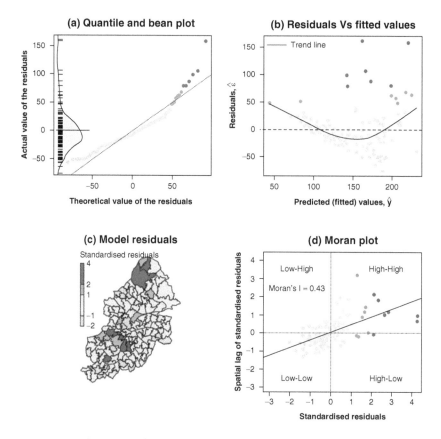

FIGURE 10.2 Visual diagnostics for the simple bivariate model

reference) and $\bar{\varepsilon}_{(u,v)}$ is the average residual for its neighbours, referred to as a spatially lagged variable. The slope of the line regressing $\bar{\varepsilon}_{(u,v)}$ against $\hat{\varepsilon}_{(u,v)}$ is Moran's I, a measure of spatial autocorrelation.[4]

Looking at the plots, we find a spatial pattern in the residuals. Locations where the model generates positive residuals tend to be surrounded by other locations where the model also generates positive residuals on average (high values are surrounded by high values: high–high). Similarly, locations where the model generates negative residuals tend to be surrounded by the same (low–low). This is evidence of like-with-like clustering: positive autocorrelation, as indicated by the

---

[4]This is a slight oversimplification. Although the slope is Moran's I in its usual form, variants of the Moran's I test are used for analysing regression residuals (see Hepple, 1998; Tiefelsdorf, 2002). Here the test is only used as a broad diagnostic. The difference would matter more if we were concerned with measuring statistical significance, although even then the change is likely to be subtle and matter only in borderline cases.

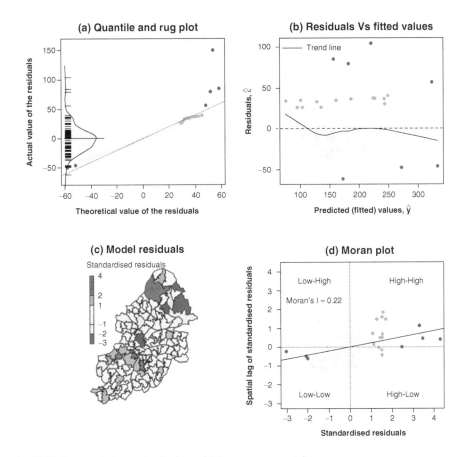

FIGURE 10.3 Visual diagnostics for the multiple regression model

upwards-sloping line. The autocorrelation reduces from the simple model ($I = 0.43$) to the multiple regression model ($I = 0.22$), indicating that the latter has explained more of the geographical patterning of the neighbourhood average house prices.

More about Moran's *I*

As well as being the slope of the line on the Moran plot, Moran's *I* can be calculated as

$$I = \frac{n}{\Sigma_i \Sigma_j w_{ij}} \frac{\Sigma_i \Sigma_j w_{ij}\left(x_i - \bar{x}\right)\left(x_j - \bar{x}\right)}{\Sigma_i\left(x_i - \bar{x}\right)^2} \tag{10.4}$$

where x is any variable to be measured for spatial autocorrelation, i and j refer to locations in the study region, n is the number of locations and w_{ij} is an entry in a spatial weights matrix, $\mathbf{W}$, indicating whether i and j are neighbours (in which case $w_{ij} > 0$) or not ($w_{ij} = 0$). Conventionally, a location is not regarded as a neighbour of itself ($w_{ii} = 0$). Although the formula may look complicated, the part $\Sigma_i \Sigma_j \, w_{ij} \left(x_i - \bar{x} \right) \left(x_j - \bar{x} \right)$ is just the covariance between the variable and the spatial lag of the variable, indicating that Moran's I is a measure of correlation between locations and their neighbours. Because of this it is usually interpreted as ranging from -1 (perfect negative spatial autocorrelation, a pattern of opposites: the black-and-white pattern on a checkerboard), through zero (no spatial autocorrelation), to $+1$ (perfect positive spatial autocorrelation, a pattern of spatial clustering: a location is the same as its neighbours). However, this is not actually correct as the range varies with the weights matrix. For the current matrix it ranges from -0.64 to $+1.02$.[5]

It is common for weights to be row-standardised; that is, for the weights of the neighbours to sum to 1 for each location. A location with two neighbours would assign a weight of 0.5 to each of them. A location with four would assign a weight of 0.25. This prevents a location with many neighbours from dominating the calculation. With the weights row-standardised, equation (10.4) simplifies to

$$I = \frac{\Sigma_i \Sigma_j \, w_{ij} \left(x_i - \bar{x} \right) \left(x_j - \bar{x} \right)}{\Sigma_i \left(x_i - \bar{x} \right)^2} \qquad (10.5)$$

and, for the regression residuals,

$$I = \frac{\varepsilon^T \mathbf{W} \varepsilon}{\varepsilon^T \varepsilon} \qquad (10.6)$$

The presence of the spatial weights matrix in the formula means that the value of Moran's I depends on the weights, which follows from how neighbours are defined. In Figures 10.2 and 10.3 locations are considered to be neighbours if they are contiguous, that is, if they share a border. Therefore the spatial weights matrix is an adjacency matrix for first-order neighbours, but second-order neighbours could also be considered. Those are neighbours of neighbours. Third-order neighbours are neighbours of the neighbours of neighbours, and so forth. Alternatively, neighbours could be places within a threshold distance of each other's centre or boundary, or we might say that neighbours are the k closest locations regardless of distance (the 10 closest, for example). Distance is an example of a symmetrical relationship (if A is within 1000 metres of B, then B is within 1000 metres of A)

---

[5]Brunsdon and Comber (2015: 232) provide the code to calculate this in R, after de Jong et al. (1984).

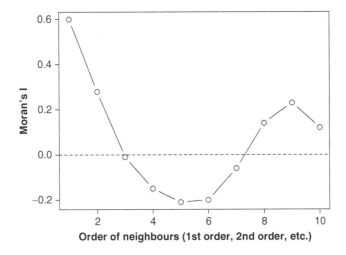

FIGURE 10.4   Showing how the measure of spatial autocorrelation is scale-dependent: patterns of similarity and difference can coexist in a study region

but k nearest neighbours is not (A can be one of the 10 nearest neighbours to B without B being one of the 10 nearest neighbours to A).

It is not surprising to find that the regression residuals exhibit a spatial patterning. This extends from the spatial patterning of the neighbourhood average house prices that are not yet fully explained by the model. Those prices (the Y variable) have a Moran's I of 0.60 based on a row-standardised spatial weights matrix of first-order neighbours. The Moran's I comparing Y with the second-order neighbours is 0.28, and by the fourth-order neighbours it is clearly negative ($I = -0.15$) before becoming positive again at the eighth-order neighbours, as shown in Figure 10.4. It follows that patterns of spatial autocorrelation are scale-dependent. Looking back at the map of average house prices in Figure 10.1, we see patterns of similarity (spatial clustering: positive autocorrelation) between places that are close together, as well as patterns of difference (spatial heterogeneity: negative autocorrelation) between places that are further apart.

10.5 PARTIAL REGRESSION PLOTS AND COLLINEARITY

The idea of multiple regression is to estimate the effect of each X variable upon the Y, net of all the other X variables. To make this clearer, imagine we are interested in exploring the relationship between the neighbourhood average house price and crimes per thousand population. We could produce a scatter plot with the average house price on the Y-axis and the crime rate on the X-axis, but that plot risks being deceptive. The problem is that the crime rate is itself correlated with the other X variables in the model: for example, the correlation between the crime rate and the percentage of properties sold that are flats and apartments is $r = 0.63$; the correlation between the crime rate and the percentage sold in 2014 is $r = 0.22$. Consequently, the scatter plot would show not the relationship between Y and X alone but the relationship with all the other X variables 'thrown in'.

To address this, we take out the effect of all those other X variables from Y and also from the X we are interested in, which here is the crime rate. If we denote the crime rate by x_8 (consistent with the variables in Table 10.1) and the other variables $x_1, \ldots, x_7$ and x_9, then we can regress Y against $x_1, \ldots, x_7$ and x_9 and calculate the residuals. Those residuals are what's left in Y once $x_1, \ldots, x_7$ and x_9 are taken out. 'What's left' includes x_8. Next, if we regress x_8 against $x_1, \ldots, x_7$ and x_9 and calculate the residuals, then what's left from that regression is the component of x_8 that is unrelated to $x_1, \ldots, x_7$ and x_9. Finally, we plot the two sets of residuals against each other to show the relationship between Y and x_8, having controlled for the other variables.

The result of this process is called a partial regression plot and it is shown in Figure 10.5 (Dunn, 1989). Note that the line of best fit has a slope exactly equal to $\hat{\beta}_8$ in Table 10.1. It confirms that what multiple regression does is estimate the effect of an X variable on Y having controlled for the other X variables in the model. The partial regression plot, which can be repeated for every X variable in turn, is useful to detect outliers, including leverage points, and also as a check on what is being fitted to the data. Looking at Figure 10.5, there are some observations

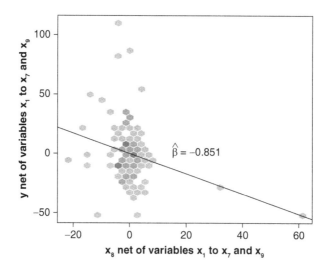

FIGURE 10.5 A partial regression plot exploring the relationship between neighbourhood average house price and the crime rate

that are apart from the rest and it is not clear that the line of best fit (and therefore $\hat{\beta}_8$ in Table 10.1) is sensible for the data.

Collinearity

The situation of the X variables being related to each other (and not just the Y variable) is called collinearity, or sometimes multicollinearity. Imagine we wanted to refit the model shown in Table 10.1 but, in addition to the percentage of properties flats or apartments, the percentage detached and the percentage terraced, we also sought to include a fourth property variable, the percentage semi-detached. Attempting to fit the model will generate an error. The reason is that all properties are flats/apartments, detached, semi-detached or terraced, so the percentages must sum to 100. That means that if we know the percentage of properties sold that were flats/apartments, the percentage detached and the percentage terraced then we already know the percentage semi-detached: it is 100 minus the sum of the rest. Not only does the percentage semi-detached variable not add any new information, it is impossible to estimate its effect on the neighbourhood average property price: it can't be disentangled from the rest.[6] This is the case of perfect multicollinearity, when one of the X variables can be predicted from one or more of the others.[7]

In practice, most X variables will be only partly correlated with each other. Even so, when the correlations become high, the estimates of the beta values become less certain (because there is, in effect, less information available on which to base the estimate), attracting larger standard errors. The variance inflation factor (VIF) is used to judge the severity of the problem.[8] As a rule of thumb, cut-offs of 5 or 10 are taken to indicate a problem of collinearity, which is not true of any of the X variables in Table 10.1. If it is found, then possible solutions are to ignore it, drop one or more of the collinear variables, collect more data (if that is possible, in the hope it will better differentiate the variables) or use a method such as factor or principal components analysis that aim to reduce sets of related variables to their core and uncorrelated parts.

---

[6] Think of it in terms of the process of creating the partial regression plot. If one X variable is fully explained by one or more of the others then there is nothing left in that X once the others have been taken out of it and you'd be left trying to explain the Y in relation to nothing.

[7] When one X variable is a linear function of one or more of the other X variables it becomes impossible to invert the matrix in the equation $\hat{\beta} = \left(\mathbf{X}^{\mathsf{T}} \mathbf{X} \right)^{-1} \mathbf{X}^{\mathsf{T}} \mathbf{y}$.

[8] The VIF value for the crime rate, for example, can be calculated by regressing x_8 against all the other predictor variables, $x_1, \ldots, x_7$ and x_9, taking the R^2 value (which shows how much of the variation in x_8 is explained by the other variables) and calculating $VIF = 1 / \left(1 - R^2 \right)$.

10.6 ALTERNATIVE MODEL SPECIFICATIONS

Mean-centring the X variables

The Y-intercept, $\hat{\beta}_0$, predicts the value of Y when all the Xs are equal to zero. In other words, it predicts the neighbourhood average house price for a neighbourhood with zero unemployment, zero detached houses, zero new builds, zero crime, zero distance to the city, and zero everything else. This is an entirely hypothetical neighbourhood and not a very realistic one. It would be more meaningful to estimate the intercept for an average neighbourhood – one with average unemployment, average percentage of detached houses, average percentage of new builds, and so forth. To achieve this is simple: it requires deducting from each X variable the mean of that variable, and using what are now mean-centred variables as replacements for the original X variables in the regression model: $\mathbf{x} \Rightarrow \mathbf{x} - \bar{\mathbf{x}}$. Doing this produces the results shown in Table 10.3. Nothing has changed from Table 10.1, except for the estimate of the Y-intercept, $\hat{\beta}_0$.

Standardizing the X variables

The mean-centring could be extended by converting each of the X variables to z-values $(\mathbf{x} \Rightarrow (\mathbf{x} - \bar{\mathbf{x}})/s_x$; Section 5.5). This will not change any of the t- or

TABLE 10.3 Regression output for the multiple regression model with mean-centred X variables

		estimate	se	t-value	p-value	95% CI
β_0	(Intercept)	148	2.17	68.2	< 0.001	[144, 152]
β_1	% Unemp	−10.2	1.21	−8.41	< 0.001	[−12.6, −7.81]
β_2	% Detached	3.12	0.272	11.5	< 0.001	[2.58, 3.66]
β_3	% Flat or apartment	0.790	0.189	4.19	< 0.001	[0.417, 1.16]
β_4	% Terraced	0.458	0.166	2.77	0.007	[0.130, 0.787]
β_5	% New build	0.134	0.225	0.595	0.553	[−0.312, 0.580]
β_6	% Sold 2013	0.566	0.622	0.909	0.365	[−0.666, 1.80]
β_7	% Sold 2014	0.684	0.524	1.31	0.195	[−0.354, 1.72]
β_8	Crime rate	−0.851	0.285	−2.99	0.003	[−1.41, −0.288]
β_9	Distance to city	−3.30	1.09	−3.03	0.003	[−5.45, −1.14]
		Residual standard error: 24.9 on 122 degrees of freedom				
	R^2	0.808		Adjusted R^2	0.794	
		F-statistic: 57.0 on 9 and 122 degrees of freedom, p-value: < 0.001				

p-values, the R^2 values or the F-statistic, but it will change the beta estimates and their standard errors because the measurement units have been changed to units of standard deviation. The advantage is that the beta estimates are now more directly comparable in that each gives the expected change in the Y variable for a change of one standard deviation in the X variable. The disadvantage is that a one standard deviation change is not especially intuitive to understand.[9]

Taking the power of an *X* variable or fitting an interaction term

If we believe the impact of an X variable on Y is non-linear, then if we may wish to take powers of X and include them in the model. For example, we suspect from Chapter 9 that there is a curvilinear relationship between the neighbourhood average house price and the unemployment rate (x_1) then that could be modelled using a polynomial relationship of the form $\hat{y} = \hat{\beta}_0 + \hat{\beta}_1 x_1 + \hat{\beta}_2 x_1^2$ (equation (9.21)).

We may also believe that two or more of the variables have a joint effect upon the Y that is in addition to their own individual effects. If so, then we may want to include an interaction term, which is the product of the X variables concerned. For example, we may wish to multiply the percentage of properties sold that are flats/apartments (x_3) by the crime rate (x_8) on the basis that flats in areas of higher crime rates are qualitatively different from those in lower crime rates – the former could be student properties, for example, whereas the second might be second homes near the financial and business centres that are occupied by business people during the working week.

Table 10.4 shows the results of adding the variables x_1^2 and $x_3 x_8$ to the model previously shown in Table 10.1. Neither of the two additional variables is significant at a conventional level of $p < 0.05$, although the interaction term $(x_3 x_8)$ almost reaches the less conservative level of $p < 0.10$. As such, it may be interesting to note that although a higher percentage of flats and apartments sold tends to increase the neighbourhood average house price, where that higher percentage coincides with a higher crime rate, it tends to decrease the average over and above the decrease expected from the crime rate alone. However, we should probably not read too much into this: the adjusted R^2 value has increased only fractionally (from 0.794 to 0.796) and an analysis of variance/F-test suggests there is little reason to prefer the more complex model over the previous one.

---

[9] The X variables could also be range-standardised, whereby $\mathbf{x} \Rightarrow \mathbf{x} / (x_{max} - x_{min}) \times 100$. This would leave the beta values estimating the expected impact on Y of a one-percentage-point increase along the range of x, which is even harder to explain.

TABLE 10.4 Regression output for the multiple regression model, now including the square of one of the X variables and also an interaction term. The X variables are mean-centred

		estimate	se	t-value	p-value	95% CI
β_0	(Intercept)	145	3.11	46.7	< 0.001	[139, 151]
β_1	% Unemp	−10.3	1.22	−8.43	< 0.001	[−12.7, −7.86]
β_2	(% Unemp)^2	0.294	0.319	0.922	0.358	[−0.337, 0.925]
β_3	% Detached	2.91	0.308	9.46	< 0.001	[2.30, 3.52]
β_4	% Flat or apartment	0.784	0.189	4.16	< 0.001	[0.411, 1.16]
β_5	% Terraced	0.375	0.173	2.17	0.032	[0.033, 0.717]
β_6	% New build	0.110	0.227	0.485	0.628	[−0.339, 0.559]
β_7	% Sold 2013	0.671	0.627	1.07	0.286	[−0.570, 1.91]
β_8	% Sold 2014	0.807	0.534	1.51	0.133	[−0.250, 1.86]
β_9	Crime rate	−1.02	0.300	−3.41	< 0.001	[−1.62, −0.430]
β_{10}	% Flat or apartment × Crime rate	−0.082	0.051	−1.61	0.111	[−0.182, 0.019]
β_{11}	Distance to city	−3.15	1.09	−2.90	0.004	[−5.30, −1.00]
		Residual standard error: 24.7 on 120 degrees of freedom				
	R^2	0.813		Adjusted R^2	0.796	
		F-statistic: 47.4 on 11 and 120 degrees of freedom, p-value: < 0.001				

Taking the logarithm of the Y variable

An alternative to transforming one or more of the X variables is to transform the Y variable.[10] Taking the logarithm, for example, may address the lingering presence of heteroscedasticity associated with the underprediction of the areas with highest average house price. The model would therefore take the form

$$\log \hat{y} = \hat{\beta}_0 + \hat{\beta}_1 x_1 + \hat{\beta}_2 x_2 + \cdots + \hat{\beta}_9 x_9$$

which (from Section 4.6) can also be written as

---

[10]What is known as the Box–Cox method is sometimes used to find a transformation of the Y variable that best fits the data. Without going into detail, the problem with this more automated approach is that it can recommend transformations that are hard to give meaning to.

$$y = e^{\hat{\beta}_0} e^{\hat{\beta}_1 x_1} e^{\hat{\beta}_2 x_2} e^{\hat{\beta}_3 x_3} \dots e^{\hat{\beta}_9 x_9} \qquad (10.7)$$

Equation (10.7) reveals that an increase in one of the X variables will have a multiplicative effect on the Y. Because of this, the interpretation of the beta estimates changes. Without taking the logarithm of Y, $\hat{\beta}_1 = -10.2$ (see Table 10.1), which means that for every one-unit increase in the neighbourhood unemployment rate, we expect to subtract 10.2 thousand pounds from the average house price. Now, fitting the model to the logarithm of Y, $\hat{\beta}_1 = -7.09 \times 10^{-2}$ (see Table 10.5), which means that for every one-unit increase in the unemployment rate we expect to *multiply* the average house price by $e^{-0.0709}$, which is 0.932. Noting that this is approximately equal to a reduction of 7 per cent ($1 - 0.932 = 0.068$) leads to a shortcut interpretation of the beta values when Y is on a log scale: they are approximately equal to the percentage change expected in Y given a one-unit change in the X.[11,12]

Comparing Table 10.5 with Table 10.3, there is an improvement to the model fit, as indicated by the R^2 values. The AIC value for the new model is -137, compared to 1235 before (recall that lower is better for the AIC measure). A Breusch–Pagan test indicates a reduction in the heteroscedasticity, but the spatial autocorrelation in the residuals has increased from $I = 0.223$ to $I = 0.271$ using a first-order neighbours weights matrix.

Fitting a model with a discrete Y variable

The neighbourhood average house price is a continuous variable that could take on any value within a given range (between zero and infinity, in principle). Sometimes the Y variable will not be continuous, in which case a different approach may be required.

Consider, for example, a data set where each row represents a pupil at a secondary (high) school and the columns are whether the pupil attends one of their nearest three schools (the Y variable), attributes of the pupil including their gender, whether they are eligible for a free school meal, their ethnicity, and attributes of the school they attend, including whether it is a single-sex school, whether it is academically selective and whether it is a school affiliated to a faith group (the X variables). We could try and fit an OLS multiple regression model of the form $\hat{y} = \hat{\beta}_0 + \hat{\beta}_1 x_1 + \hat{\beta}_2 x_2 + \cdots$, but doing so encounters problems because of the binary nature of the Y variable whereby a pupil either attends one of their three nearest schools ($y = 1$) or they don't ($y = 0$). The residuals will not be normal but will be

---

[11] See, for example, Gujarati (2011). Faraway (2015: 134) notes that the approximation is quite good up to about ±0.25.

[12] If both the X and Y variables are logged then the beta values can be interpreted as the percentage change expected in Y given a percentage change in the X.

TABLE 10.5   Regression output for the multiple regression model with the logarithm of the original Y variable. The X variables are mean-centred

		estimate	se	t-value	p-value	95% CI
β_0	(Intercept)	4.94	1.20×10^{-2}	412	< 0.001	[4.91, 4.96]
β_1	% Unemp	-7.09×10^{-2}	6.73×10^{-3}	-10.5	< 0.001	[−0.084, −0.058]
β_2	% Detached	1.49×10^{-2}	1.51×10^{-3}	9.89	< 0.001	[0.012, 0.018]
β_3	% Flat or apartment	3.57×10^{-3}	1.05×10^{-3}	3.42	< 0.001	[0.002, 0.006]
β_4	% Terraced	1.64×10^{-3}	9.19×10^{-4}	1.78	0.078	[−0.000, 0.003]
β_5	% New build	-5.69×10^{-4}	1.25×10^{-3}	-0.456	0.649	[−0.003, 0.002]
β_6	% Sold 2013	4.16×10^{-3}	3.45×10^{-3}	1.21	0.230	[−0.003, 0.011]
β_7	% Sold 2014	3.73×10^{-3}	2.91×10^{-3}	1.28	0.202	[−0.002, 0.010]
β_8	Crimes p.p.	-5.10×10^{-3}	1.58×10^{-3}	-3.23	0.002	[−0.008, −0.002]
β_9	Distance to city	-2.07×10^{-2}	6.03×10^{-3}	-3.44	< 0.001	[−0.033, −0.009]
		Residual standard error: 0.138 on 122 degrees of freedom				
R^2		0.827		Adjusted R^2	0.814	
		F-statistic: 64.8 on 9 and 122 degrees of freedom, p-value: < 0.001				

heteroscedastic, and the predicted values could be out of range and nonsensical – values less than zero or greater than one.

For the sake of argument, let's call attending a nearest school a success, whereas not doing so is a failure. A common approach for handling binary outcomes is to fit a model of the form,

$$\hat{y} = \log\left(\frac{p}{1-p}\right) = \hat{\beta}_0 + \hat{\beta}_1 x_1 + \hat{\beta}_2 x_2 + \cdots \tag{10.8}$$

where p is the probability of success (attending a nearest school), $1 - p$ is the probability of failure, $p/(1-p)$ is the ratio of success to failure, giving the odds ratio (of success), and the log of that ratio is called the log-odds. The model is known as a logit model (pronounced 'lo-jit', but more identifiable as a weak pun when written as log-it) and although it is not estimated using ordinary least squares, the principles of the actual estimation procedure are not dissimilar: to try and identify the most likely beta parameters for the data and to maximise the fit.[13]

---

[13]An alternative to the logit model is the probit model and many other types of model exist. For example, for count data where the distribution is skewed to lower numbers and higher numbers are rare, a Poisson model may be used.

Table 10.6 shows the results of such a model, for pupils in secondary schools in London in 2011. There is no R^2 given for this model, although a pseudo-R^2 could be calculated as the square of the correlation between the observed and predicted values. It is 0.10, suggesting there is still a lot of the variation in the Y value to be explained. The AIC or a similar measure of fit is preferred, but is harder to interpret. The AIC for the model is 577,538, which means little in itself but is useful for comparing models to see if an improved model fit justifies the additional variables. A positive beta estimate means the predictor variable is associated with increased success, and a negative value with decreased success. However, the estimates are not entirely intuitive because they are measured on the logit scale.

TABLE 10.6 *The results of a logit model looking at which groups of pupils in London attended one of their three nearest secondary schools*

		estimate	se	z-value[1]	p-value	$e^{\hat{\beta}}$ (odds ratio)	95% CI (as an odds ratio)
β_0	(Intercept)	0.434	0.008	57.0	<0.001		
β_1	Female (base=male)	0.105	0.007	14.5	<0.001	1.11	[1.10, 1.13]
β_2	Asian	0.142	0.009	16.3	<0.001	1.15	[1.13, 1.17]
β_3	Black	−0.541	0.009	−62.7	<0.001	0.582	[0.573, 0.592]
β_4	Chinese	−0.253	0.037	−6.75	<0.001	0.777	[0.722, 0.779]
β_5	Mixed	−0.273	0.012	−22.5	<0.001	0.761	[0.743, 0.002]
β_6	Other	−0.340	0.015	−23.4	<0.001	0.712	[0.692, 0.732]
β_7	Unclassified (base=White)	−0.531	0.023	−23.2	<0.001	0.588	[0.562, 0.615]
β_8	Free school meal eligible (base=ineligible)	0.011	0.008	1.50	0.134	1.01	[0.997, 1.03]
β_9	Boys' school	−0.689	0.012	−58.3	<0.001	0.502	[0.491, 0.514]
β_{10}	Girls' schools (base=mixed school)	−0.490	0.010	−51.1	<0.001	0.613	[0.602, 0.625]
β_{11}	Foundation school	0.067	0.008	8.64	<0.001	1.07	[1.05, 1.09]
β_{12}	Academy	−0.065	0.011	−6.18	<0.001	0.937	[0.918, 0.956]
β_{13}	Voluntary aided faith school	−0.989	0.009	−107	<0.001	0.372	[0.366, 0.379]
β_{14}	Voluntary controlled faith school (base=Community school)	−0.335	0.027	−12.6	<0.001	0.715	[0.679, 0.753]
β_{15}	Secondary Modern	0.044	0.018	2.41	0.016	1.04	[1.01, 1.08]
β_{16}	Selective (grammar) school (base=Comprehensive)	−1.41	0.021	−68.7	<0.001	0.243	[0.234, 0.253]

[1]The z-value replaces the t-value in this table, because the test statistic ($\hat{\beta}/se_{\hat{\beta}}$) is expected to have a normal instead of a t-distribution, although the two are very similar for a data set of this size.

As it happens, each of the predictor variables is itself a binary variable, also known as a dummy variable: a pupil is either female ($x_1 = 1$) or not ($x_1 = 0$); the pupil is either of an Asian ethnicity ($x_2 = 1$) or not ($x_2 = 0$), and so forth. Raising the exponential value to the power of these beta values ($e^{\hat{\beta}}$) gives the change in the odds of success relative to a baseline group. For example, females appear to have 1.11 (11 per cent) greater odds of attending a nearest school than males. Asian pupils are more likely to attend their nearest school than White pupils, but for Black pupils the odds are almost halved. A pupil attending a single-sex school has about half the odds of attending a nearest school compared to a pupil in a mixed school. A pupil attending a Foundation school is about 7 per cent more likely to be in one of their nearest schools than a pupil attending a Community school, but for faith schools, and especially voluntary aided faith schools (which are more likely to have affinity to the faith as part of their entrance requirements), the odds are much reduced, although not as much as they are for academically selective schools relative to comprehensives – a pupil in a selective school (which has an entrance exam) has only one quarter of the odds of attending a nearest school compared to a pupil in a comprehensive (which doesn't).

10.7 SPATIAL REGRESSION MODELS

Writing in the introduction to a book entitled *Spatial Regression Models*, the series editor, Tim F. Liao, notes:

> Much of quantitative social science methodology is about the analysis of individual behavior … most commonly in a regression type of framework. … The level of analysis … does not have to be at the individual or micro level. Sometimes researchers analyze data that are at an aggregate level, such as one representing neighborhoods, communities, firms, cities, counties, states, and nations. Still the same analytic logic applies. We seek to establish some association, causal or not, between the dependent variable and (some of) the independent variables ….When we do so, we implicitly assume that the geographical or spatial locations of the observations in the analysis do not matter. (Ward and Gleditsch, 2008: vii)

It's worth repeating that last sentence, just to emphasise it: 'we implicitly assume that the geographical or spatial locations of the observations in the analysis *do not matter*'.

Regression is not a spatial method of analysis. It takes columns of data and looks for an association between the variables they describe. What it does not do is consider where the data were observed. The locations are required to look for geographical patterns in the variables or in the residuals, but those mappings occur before or after the regression, not as part of it. The regression itself is blind to geography and somewhat unsympathetic to it. Recall the residuals are assumed to be identically *and independently* distributed, which presumes no spatial autocorrelation amongst them.

Compare that with Moran's *I*, which includes a spatial weights matrix in its calculation (see equation (10.4)). The matrix defines a geographical relationship between locations, identifying places as neighbours or not. If the places were moved, their neighbours would be altered, the matrix would need to be updated and the Moran's *I* statistic would change. Moran's *I* is an explicitly spatial statistic because it is part of a set of methods whose results change when the locations of the objects being analysed change (Longley et al., 2015: Chapter 13).

A spatial regression model also incorporates a spatial weights matrix in its specification. Ward and Gleditsch (2008) give two examples. Beginning with a standard regression of the form

$$y = \mathbf{X}\beta + \varepsilon \qquad (10.9)$$

a spatial error model is reached by dividing the error into two parts: $\varepsilon = \lambda \mathbf{W}\xi + \epsilon$. From left to right, these are a spatially correlated error term (note the weights matrix) and a spatially uncorrelated one, giving

$$y = \mathbf{X}\beta + \lambda \mathbf{W}\xi + \epsilon \qquad (10.10)$$

where λ (lambda) is a measure of spatial autocorrelation, indicating the extent to which neighbouring residuals are correlated. Applying this to the model previously considered in Table 10.4 yields $\lambda = 0.709$, revealing a high level of positive spatial autocorrelation amongst the residuals. Ignoring that spatial autocorrelation, as the former model did, will tend to leave the standard errors of the beta values underestimated. This may not matter, especially if we are not interested in the statistical significance of those values, except that the beta estimates may themselves become 'unstable' and be further from their true value. More particularly, to ignore the spatial autocorrelation is to ignore the processes that caused it – that is, to ignore the most interesting geographical questions.

The spatial error model says little about those processes other than suggesting they exist: 'in a spatial error model specification, the observations are related only due to unmeasured factors that, for some unknown reason, are correlated across the distances among the observations' (Ward and Gleditsch, 2008: 67). In contrast, a spatially lagged *Y* model formalises a distinct process whereby the outcome observed in one place affects and is affected by the outcomes in its neighbours. For example, if processes of gentrification or urban renewal raise the average house price in one neighbourhood, rather than assuming the effect of those processes somehow stops at the boundary of the neighbourhood, it is instead modelled with the possibility of overspill into surrounding neighbourhoods. This sort of process has been shown with the reopening of businesses in the aftermath of Hurricane Katrina.[14]

Starting again with the standard regression model in equation (10.9), a spatially lagged Y model is obtained by dividing the error into the spatial lag of the Y variable and an independent error term,

$$y = X\beta + \rho Wy + \epsilon \tag{10.11}$$

where Wy is the spatially lagged Y variable and ρ (rho) is the spatial autocorrelation term. The presence of Y on both sides of the equation complicates the estimation of the beta values. It also complicates their interpretation. Normally a beta value is the expected change in the Y variable for a change in an X variable, holding all the other Xs constant. For a logged Y variable, it can be interpreted as a percentage change. In Table 10.5, the value of $\hat{\beta}_1 = -7.09 \times 10^{-2}$ (i.e. −0.0709) suggests that a one-unit increase in the unemployment rate will lead to a 7 per cent decrease in the neighbourhood average house price. With the spatially lagged Y model the estimated value is $\hat{\beta}_1 = -5.43 \times 10^{-2}$, which implies it will be about 5 per cent or, more precisely, 5.28 per cent (from $e^{-0.0543}$). However, that interpretation is not correct because there is now the spatial spillover to consider. Rearranging equation (10.11), we obtain

$$y - \rho Wy = X\beta + \epsilon \text{ or, equivalently, } (I - \rho W)y = X\beta + \epsilon$$

which means that

$$E(y) = (I - \rho W)^{-1} X\beta \tag{10.12}$$

where $E(y)$ is the expected value of y and $(I - \rho W)^{-1}$ is the spatial spillover or multiplier. Because of the neighbourhood relations, an increase in the X variable at one or more locations will have a knock-on effect in neighbouring locations, rippling out to their neighbours, and on to other locations, as well as back on to the locations that initiated the process, amplifying or attenuating its effect. For example, Figure 10.6 maps the predicted decrease in the neighbourhood average house price for a one-unit increase in unemployment at location A and, separately,

---

[14]See LeSage *et al.* (2011). Although theirs is a probit model, modelling the probability a business reopens, the idea that one business's decision affects others on the same street, and vice versa, is an example of a spatial spillover.

at location B. The effect of the increase diminishes with increased distance from A or B. The estimated decrease in the average house price at A is 5.47 per cent at equilibrium (once the spatial multiplier has been accounted for). At B it is 5.46 per cent. Both are higher than the estimate obtained from the beta value alone (5.28 per cent). One way to interpret the beta values is to ask what would happen if there were a one-unit change in an X value at all locations simultaneously, which can be calculated as $\hat{\beta}_k / (1 - \rho)$. For a region-wide, one-unit increase in the unemployment rate, where $\hat{\beta}_1 = -5.43 \times 10^{-2}$ and $\rho = 0.399$, the total effect is a decrease in the neighbourhood average house price of about 9 per cent.

Anselin and Rey (2014) provide a comprehensive introduction to these and other spatial regression models, a decision tree to help choose between them, information about how they are estimated and a tutorial in the various software available to do so (see also Anselin, 2005). Purely on the basis of fit, and the AIC measure of it, the spatially lagged Y model (AIC = −173) is better than the spatial error model (AIC = −168) and both outperform the OLS model (AIC = −137) for the neighbourhood house price average. Recall that a lower AIC value indicates a better fit. However, there still is a lot of spatial autocorrelation in the residuals from the spatially lagged Y model. An even better-fitting model incorporates the spatial lag of both the Y and the error term, $\mathbf{y} = \mathbf{X}\boldsymbol{\beta} + \rho\mathbf{W}\mathbf{y} + \lambda\mathbf{W}\boldsymbol{\xi} + \boldsymbol{\epsilon}$ (AIC = −178). This

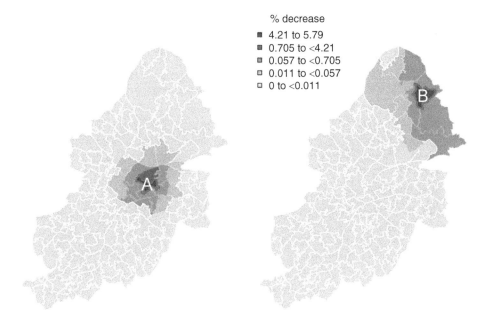

FIGURE 10.6   Showing the spatial multiplier: the expected decrease in the neighbourhood average house price as a consequence of a one-unit increase in the unemployment rate at A or B

takes us back to asking *why* the residuals are spatially correlated and to considering whether the problem lies in the specification of the model. In particular, are there other predictor variables that should be included to more fully explain the house price averages, or perhaps the property market in Birmingham consists of a number of distinct spatial regimes that cannot be well understood by fitting a single model to them all? Maybe we need to adjust our focus to look for geographical variation and difference.

10.8 CONCLUSION

As suggested at the beginning of the chapter, there is much more that can be written about regression methods, and many textbooks available that have done so. For example, we could consider the case where the Y is neither continuous nor binary but is used to indicate which of a set of groups or categories each of the observations belongs to.[15] We could give consideration to the situation where one or more of the X variables is correlated with the residuals and the possibility of using what are called instrumental variables to solve it,[16] what to do when there are missing data,[17] or how to better establish whether some variables cause a change in another rather that simply being associated with it.[18]

Instead, however, we will return to the geographical focus of this chapter and consider a set of steps that will be useful when analysing geographical data. First, map the data and look for evidence of spatial clustering and/or heterogeneity. Next, fit the proposed regression model, do the more standard checks of the model, but then also map and check the residuals for evidence of spatial autocorrelation. If the evidence is there, and there is little reason to suppose that the model has been badly misspecified, then consider a more geographical approach, which may include the use of spatial regression models.

Even so, the possibility will remain that it is overly simplistic to try and fit a single model to try and capture the geographical complexity of a study region and the varying contexts within which the outcomes of interest are generated. What's

---

[15]Multinomial regression.

[16]The idea is to try and replace the problematic X variable with predictions of it from another variable where those predictions are no longer correlated with the residuals.

[17]Leave them out, replace the missing values with the average of the variables concerned or use data imputation methods to try and predict the missing value from other information about the observation, including possibly its location.

[18]Establishing exact causes is really hard and usually requires a randomised experiment or a situation where two sets of observations are expected to be alike in almost every way except only one was exposed to or follows on from some sort of intervention such as a change in government policy. In such a circumstance it is reasonable to suppose that any consistent difference between the sets is due to that intervention.

produced by a regression model may be akin to the stylised facts described at the beginning of this book (in the introduction to Chapter 1), acting to average away the differences between places rather than stopping to see them as something of geographical interest. It may be that a more geographically nuanced approach will be needed that searches for and recognises local patterns and relationships, emphasising spatial differences rather than trying to force things into a one-size-fits-all sort of model. Such approaches include local indicators of spatial association, spatially hierarchical models and geographically weighed regression, each of which is discussed in the following chapter.

11

ANALYSING GEOGRAPHICAL PATTERNS AND DIFFERENCES

11.1 INTRODUCTION

One of the criticisms of quantitative geography has been that it is nomothetic – looking for general rules of, say, human behaviour and insensitive to or not interested in variation or difference. The focus, it is argued, is on average outcomes and relationships. As Chapter 1 discussed, it is true that some of the origins of quantitative geography lie in a period during the 1960s and 1970s when theories and 'laws' of economics, as well as physics, acted as inspiration for testing or modifying those ideas in the spatial domain (see Johnston and Sidaway, 2015). However, as a caricature of what takes place under the umbrella of contemporary quantitative geography, the criticism is anachronistic. With greater computational power and more geographically detailed data sets the approach is often idiographic – looking for geographically distinct events and outcomes, searching for geographical differences and spatial variations, and seeking to measure 'things' in the geographical context that helped to produce them. The focus is less on averages than on geographical deviations from those averages.

This chapter considers some distinctly geographical methods of analysis, focusing on the analysis of point patterns, local indicators of spatial association, geographically weighted statistics and some methods of analysis that fall with a regression framework but outside the spatial econometric paradigm discussed in Section 10.7. Of all the chapters in this book, this one is especially focused on the geography in quantitative geography.

11.2 SPATIAL CLUSTERING AND CONCENTRATION PROFILES

The information shown in Table 11.1 concerns children of secondary (high) school age who attended a state-funded school in London in 2011. It shows the proportion of pupils living within 5 km and 10 km of each other's residential postcode, calculated separately for each of six ethnic groups. For these pupils, 30 per cent of Bangladeshi children live within 5 km of each other and a clear majority (61 per cent) live within 10 km. In contrast, only 5 per cent of the White British pupils live within 5 km of each other, and 17 per cent within 10 km.

TABLE 11.1   Showing the proportion (and percentage) of each ethnic group living within 5 km and 10 km of each other

Ethnic group	Within 5 km	Within 10 km
Asian Bangladeshi	0.300 (30.0%)	0.608 (60.8%)
Asian Pakistani	0.150 (15.0%)	0.313 (31.3%)
Asian Indian	0.130 (13.0%)	0.286 (28.6%)
Black Caribbean	0.106 (10.6%)	0.309 (30.9%)
Black African	0.085 (8.5%)	0.275 (27.5%)
White British	0.054 (5.4%)	0.173 (17.3%)

Figure 11.1 provides the same information in pictorial form and extends it to include the proportions at other threshold distances (intervals of 0.5 km from 0.5 to 10 km). A steeper and faster-rising curve from the bottom left of the graph shows that a greater proportion of the group is living at shorter distances from each other. This implies that the Bangladeshi pupils are the most spatially clustered of the groups and the White British the least so. In part this is because there are so many more White British pupils in London that it would be surprising if they were not spread out: whereas 34 per cent of all the pupils are White British, only 5 per cent are Bangladeshi. However, it is not simply that more people means less clustering. For example, there are more Bangladeshi than Pakistani pupils, but the former appear more clustered than the latter.

Figure 11.1 is sometimes known as a concentration profile. As well as allowing for comparison between the ethnic groups, it shows the proportions that

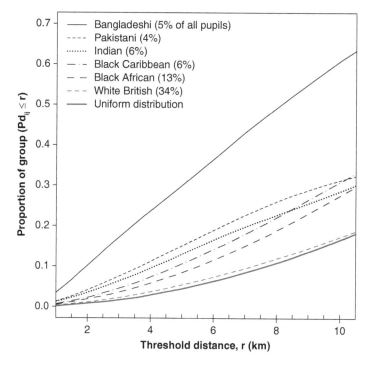

FIGURE 11.1 Concentration profile showing the proportion of London pupils living within given distances of each other

would arise if pupils were uniformly spaced across the study region (if they were evenly spread out).[1] With the exception of the White British, the groups depart noticeably from that benchmark. It means that if we map the places where the pupils live, we should expect to find there are locations where a given ethnic group is more noticeably prevalent and places where it is not. This is confirmed in the maps of Figure 11.2, which uses kernel density estimation to emphasise the places where the groups are most spatially concentrated. The clustering of the Bangladeshi group is clearly evident, as is the much greater spreading out of the White British.

Figure 11.3 shows another way of constructing a concentration profile, suitable for area data. In this example the areas are raster cells, although they might instead be Census neighbourhoods or other administrative zones. What the profile shows

---

[1]For most of the study region and for smaller values of r the expected proportion is $\pi r^2/A$, where A is the area of the study region. However, this is not the case at the edge of the region where a circle located by the edge and of radius r would extend outside the study area. Therefore an edge correction is required.

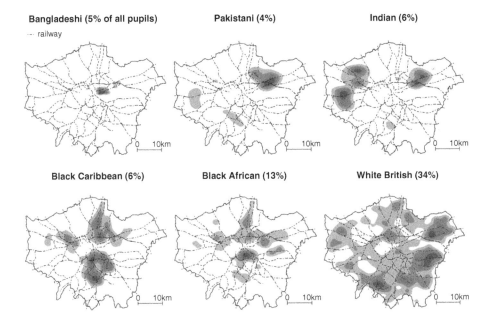

FIGURE 11.2 Using kernel density estimation to map the places where school children of each ethnic group are most prevalent in London

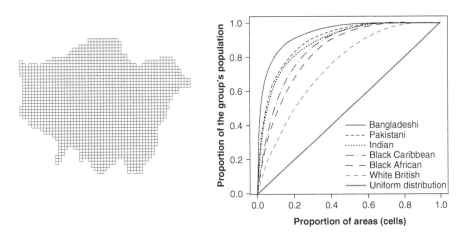

FIGURE 11.3 A second concentration profile showing what proportion of each ethnic group's school age population can be found in a given proportion of the areas shown on the left

is that 90 per cent of all the Bangladeshi pupils reside in less than 20 per cent of the areas, whereas, for example, the White British are more spread out: 90 per cent reside in 57 per cent of the areas.

11.3 POINT PATTERN ANALYSIS AND COMPLETE SPATIAL RANDOMNESS

In their book, *Geographic Information Analysis*, David O'Sullivan and David Unwin (2010) discuss various ways of undertaking point pattern analysis. At the heart of many of theses methods is a comparison of the pattern as it has occurred with what would be expected under a process of complete spatial randomness – if the locations of events were chosen randomly within the study region.

Figure 11.4 maps the locations of crimes that took place in Baton Rouge, Louisiana, in 2014. The data are openly available, as they are for many other cities.[2] Looking at the patterns of homicide, they don't appear to be random, but what would a random pattern look like? It is tempting to assume that a random pattern would lack any evidence of spatial clustering anywhere across the map. However, that cannot be true: if the process is always constrained so that its outcomes never

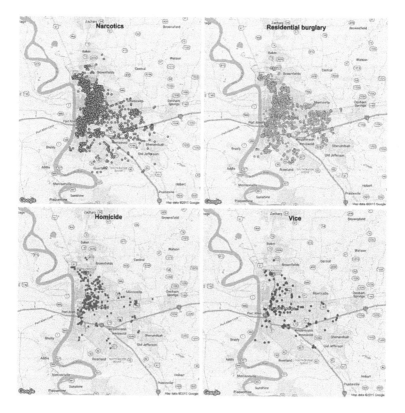

FIGURE 11.4   Incidents of crime in Baton Rouge, LA, in 2014

---

[2]You can obtain them at the US City Open Data Census: http://us-city.census.okfn.org

cluster together, then that constraint means the process cannot be random. Spatial patterns emerge even from random events.

One way to test the statistical significance of the point pattern is to overlay the points with a raster grid (a quadrat) and count the number of occurrences in each grid cell. These are the observed values and they are shown for the homicides data in Figure 11.5. The expected value is constant for each grid cell and is simply the number of homicides that took place in 2014 divided by the number of grid cells. This is the intensity of the process, λ, and in this instance is equal to 293/110, which is 2.66 – that is, there were an average of 2.66 murders per grid cell in 2014. Having the observed and expected values per grid cell, a chi-square test (χ^2) can be calculated (see Chapter 6) with degrees of freedom equal to the number of grid cells minus one. For the homicide data, this gives the highly statistically significant value of $\chi^2 = 1643$ ($p < 0.001$), and so the null hypothesis of complete spatial randomness is rejected.

A problem with this approach is that the chi-square test requires the expected value to be greater than 5, which it isn't. One solution is to increase the size of the grid cells, but if we don't want to do that then another is to use a simulation approach that assigns grid cells with values that are randomly drawn from a Poisson distribution with mean equal to the intensity, λ. The chi-square value for the observed data can then be compared with the value obtained with the randomly drawn data. If the process is repeated with, for example, 100 or 1000 random draws,

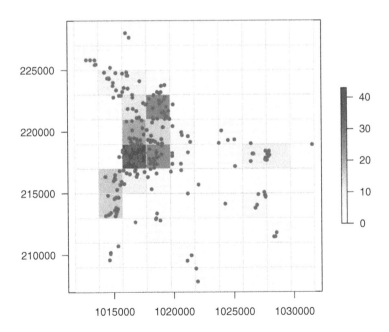

FIGURE 11.5 Quadrat analysis of the homicides data

then it is possible to assess how unusual the observed chi-square value is *vis-à-vis* the chi-square values obtained for the random replicates. For the homicide data, the hypothesis of complete spatial randomness is again rejected.

The Poisson distribution is often used to model count data and to simulate spatially random events. Its formula gives the probability of obtaining a count of x when the average intensity of the pattern is λ:

$$P(x) = \frac{\lambda^x e^{-\lambda}}{x!} \tag{11.1}$$

For $\lambda = 2.66$ the probability of no homicides occurring within a grid cell is $P(x=0) = \lambda^0 e^{-\lambda}/0! = 0.070$, for one homicide it is $P(x=1) = \lambda^1 e^{-\lambda}/1! = 0.186$, it peaks at two, $P(x=2) = 0.247$, and subsequently declines (see Figure 11.6). A feature of the Poisson distribution is that its variance is equal to its mean. That is not true of the homicide counts shown in Figure 11.5. Their variance is 40.2, which is 15 times as large as the mean (λ). This confirms that their distribution is not Poisson but exhibits a large degree of spatial clustering.

A limitation of the quadrat tests is they are sensitive to the cell size and to the positioning of the raster grid over the data. Moreover, the test gives no indication of the scale of clustering. An alternative approach is to calculate the distance from each crime event to the geographically closest occurrence of the same

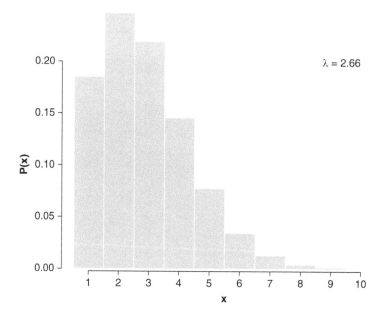

FIGURE 11.6 An example of a Poisson distribution

crime. The proportion that is within 100 metres, 200 metres, 300 metres or other threshold distances of one another can then be calculated in a manner equivalent to the concentration profile discussed earlier, the difference being that we are now considering only the nearest neighbour distance. The result is called a G-function and is shown for the crime data in Figure 11.7. What we find is that a much greater proportion of narcotics crimes occurred within a very short distance of each other – typically within 200 m of the nearest neighbour – whereas vice, for example, is more spatially diffuse. Even so, all the crimes are more spatially clustered than under a process of complete spatial randomness (the theoretical value shown). The envelope around the theoretical values give

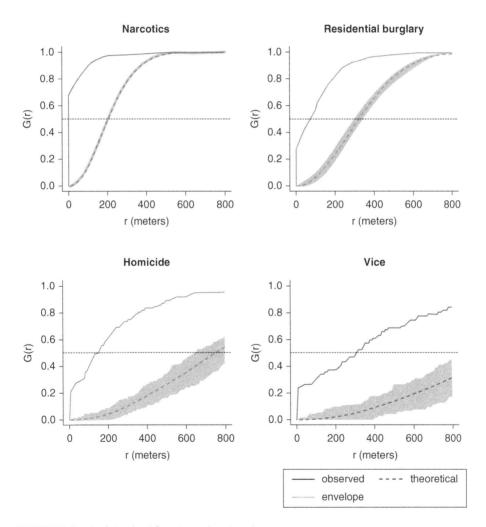

FIGURE 11.7   Applying the G-function to the crime data

the range of G-values that arise from samples of point locations drawn at random from the study region, where the number of points in each sample is equal to the number of crime events (i.e. the number of narcotics crimes in the top left profile, the number of burglaries in the top right, and so forth). If there are 100 samples and the envelope includes the range for the 'middle' 95, then what it shows is akin to a 95 per cent confidence interval. In this way the crimes are found to be significantly more clustered than random.

Whether spatial randomness is the right benchmark to judge the clustering is a moot point. Under a process of complete spatial randomness, a crime event can occur anywhere within the study region with equal probability of occurrence at all locations. In reality, that is not true. Crimes are generally more likely in areas of higher population density and not in the Mississippi River. One possibility would be to restrict the random sample from all possible locations to a more plausible subset: those at or within a short distance of where crimes have taken place over recent years, or the centres of zip codes.

The G-function is only one of a number of related functions that can be used to analyse point patterns. These include the F-function, J-function, K-function and the L-function. The F-function, for example, considers the minimum distance from random locations in the study region to an event (here an event is the occurrence of a crime), whereas the K-function extends beyond nearest neighbour distances to consider all the distances between events. For further information, see O'Sullivan and Unwin (2010), Brunsdon and Comber (2015) and Lloyd (2014).

A further way to extend the analysis is to consider time as well as geography. For example, Figure 11.8 considers the distance between the location of a crime

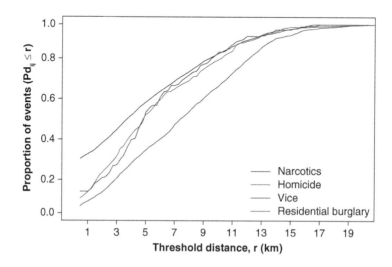

FIGURE 11.8 Showing the distance between consecutive crime events of the same type as each other

and the next incident of the same type of crime. What we find is that a higher proportion of narcotics crimes occur one after another at short distances apart, whereas residential burglaries tend to move around the city more. A simple explanation for the spatial and temporal clustering of narcotics crimes is that multiple arrests are made at the same time in a drugs bust.

11.4 LOCAL INDICATORS OF SPATIAL ASSOCIATION

Figure 11.9 is a Moran plot of the log of the neighbourhood average house prices discussed in Chapter 10, based on a first-order contiguity. In other words, it compares the expected value of a house in each Census neighbourhood with the average expected value in adjacent neighbourhoods. The correlation is positive, indicating positive spatial autocorrelation: the Moran's value is $I = 0.595$, against a theoretical range from -0.641 to 1.02 and an expected value (essentially the 'zero value') of -0.008. There are various ways to judge the statistical significance of this result. One way is to form a test statistic, $(I - E(I))/\sigma_I$, make an assumption about how that test statistic will be distributed given repeated resampling, and see where on the distribution the observed value lies.[3] For example, a normality assumption

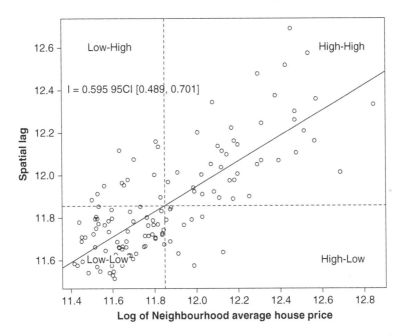

FIGURE 11.9   Moran plot of the neighbourhood average house price in Birmingham, England

---

[3] The expected value, E(I), is $-1/(n-1)$. The variance and therefore standard deviation are harder to calculate: see Brunsdon and Comber (2015) and Fotheringham (2000).

can be made, in which case a test statistic of magnitude greater than 1.96 can be judged statistically significant at the 95 per cent confidence level ($p < 0.05$) or 2.58 at 99 per cent confidence ($p < 0.01$). Another is to use a simulation approach – also known as a Monte Carlo or permutation approach – that works by shuffling the data around the locations in the study region and seeing how unusual the actual Moran's value is when compared to the values obtained from randomly relocating the data. For instance, 1000 random permutations of the neighbourhood average house prices generated a set of values, of which 99 per cent were within the range from −0.137 to 0.155. Given that the observed value of $I = 0.595$ is outside this range, we can say it is statistically significant with a pseudo p-value of $p < 0.01$.

The formula for Moran's I was given in equation (10.4):

$$I = \frac{n}{\Sigma_i \Sigma_j w_{ij}} \frac{\Sigma_i \Sigma_j w_{ij} \left(x_i - \bar{x} \right)\left(x_j - \bar{x} \right)}{\Sigma_i \left(x_i - \bar{x} \right)^2}$$

Noting that both $n / \Sigma_i \Sigma_j w_{ij}$ and $\Sigma_i \left(x_i - \bar{x} \right)^2$ are constants for a given set of data and their spatial weights, then,

$$I \propto \sum_i \sum_j w_{ij} \left(x_i - \bar{x} \right)\left(x_j - \bar{x} \right) \tag{11.2}$$

where $\propto$ means 'proportional to' (i.e. I increases with the sums). Note also that

$$\sum_i \sum_j w_{ij} \left(x_i - \bar{x} \right)\left(x_j - \bar{x} \right) = \sum_j w_{1j} \left(x_1 - \bar{x} \right)\left(x_j - \bar{x} \right) + \sum_j w_{2j} \left(x_2 - \bar{x} \right)\left(x_j - \bar{x} \right)$$

$$+ \sum_j w_{3j} \left(x_1 - \bar{x} \right)\left(x_j - \bar{x} \right) + \cdots + \sum_j w_{nj} \left(x_n - \bar{x} \right)\left(x_j - \bar{x} \right)$$

or, more simply,

$$\sum_i \sum_j w_{ij} \left(x_i - \bar{x} \right)\left(x_j - \bar{x} \right) = I_1 + I_2 + I_3 + \cdots + I_n \tag{11.3}$$

The standard Moran's statistic is a global measure of spatial autocorrelation, where 'global' means not the whole world but one value for the whole of the study region. What equation (11.3) shows is that the global value can be decomposed into a series of local Moran statistics, one for each location in the study region, where I_1 looks at the correlation between location 1 and its neighbours, I_2 looks at the correlation between location 2 and its neighbours, and so forth. These are local indicators of spatial association often known as LISAs (Anselin, 1995).

Figure 11.10 plots the log of the neighbourhood average house price for the Birmingham study area. The clusters of higher and, to a lesser extent, lower house prices that are evident in this map also appear in Figure 11.11, which shows the

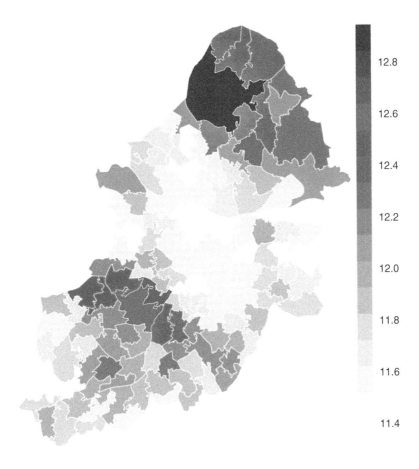

FIGURE 11.10   Map of the neighbourhood average house price on a log scale

local Moran values, calculated using a row-standardised, first-order contiguity matrix. Not all the local Moran values are statistically significant with a p-value less than 0.05. A black border emphasises those that are.

The meaning of the p-value comes into question again here, although for a reason different from those discussed in Chapter 6. The issue now is one of repeat testing – of having searched 132 times for a 'significant' result (132 is the number of neighbourhoods and the number of local Moran values they have generated). The situation is analogous to tossing a coin ten times. For each specific toss, the probability of it landing heads is $P = 0.5$, but the probability of it landing heads for one or more of the tosses is much higher, $P = 0.999$. For an unusual coin, where the probability of it landing heads is $P = 0.05$ on each toss, the probability of it landing heads for at least one of ten tosses is $P = 0.401$. However, raise it to 132 tosses and it becomes almost inevitable: $P = 0.999$. In the same way, if we are

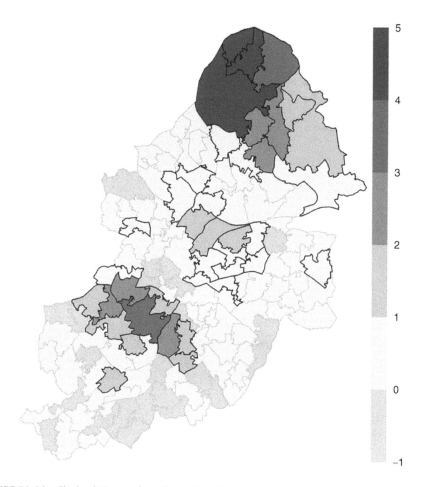

FIGURE 11.11 The local Moran values. Those with a black border might be regarded as statistically significant ($p < 0.05$), but there is no adjustment for multiple testing

willing to look for a p-value of 0.05 or less in enough places across the map then the chances are we will find it.

A traditional response to this problem is to make what is known as a Bonferroni correction, dividing the sought after p-value by the number of tests to get a replacement p-value that should be looked for instead. For example, $p < 0.05/132 = 3.79 \times 10^{-4}$ would be required for the equivalent of $p < 0.05$. The problem with this approach is that it will usually reveal few if any significant clusters and is, in any case, unnecessarily conservative given that there are not really 132 separate tests but some number less than 132, because the neighbours at each location overlap. Brunsdon and Comber discuss alternative procedures; the result of one of these, developed by Holm (1979), is shown in Figure 11.12.

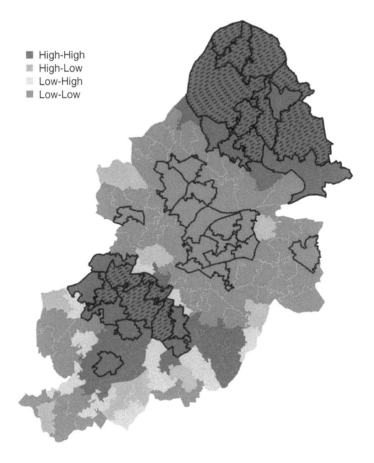

FIGURE 11.12 Showing hot spots and cold spots in the neighbourhood average house price. Those with a black border might be regarded as statistically significant ($p < 0.05$), but those with a dashed crosshatching remain so after adjustment for multiple testing

Figure 11.12 also makes a distinction between areas of high average house price with neighbours of high house price (high–high: 'hot spots') and areas of low average house price with neighbours of low house price (low–low: 'cold spots'). It is helpful to do so because in terms of their local Moran values they are equivalent: both high–high and low–low occurrences generate positive Moran values (both are examples of like-with-like clustering, of positive spatial autocorrelation). The map also shows examples of negative spatial autocorrelation, and therefore negative Moran values, where either high values are surrounded by low ones, or low values by high ones.

Another local statistic that can also be used to highlight patterns of spatial clustering is the G★-statistic, which can be expressed as

$$G^{\star}_{(u_i,v_i)} = \frac{\sum_{j=1}^{n} w_{ij} x_j}{\sum_{j=1}^{n} x_j}$$

(11.4)

where $G^{\star}_{(u_i,v_i)}$ is the local statistic, calculated at location i, which has grid coordinates (u, v) and for which a binary weights matrix is used, set to $w_{ij} = 1$ if locations i and j are within a certain distance, d, of each other or else $w_{ij} = 0$. The weights are not row-standardised (Ord and Getis, 1995). For the homicide data, what we will be calculating is the proportion of all homicides that are

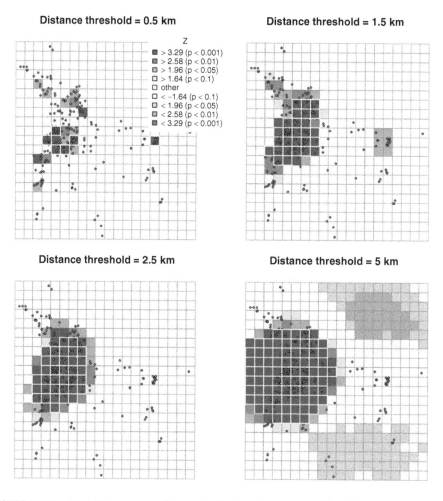

FIGURE 11.13 G*-statistics, expressed as standardised z-values, calculated for the homicide data and at varying distance thresholds. Cells shown with a yellow border are considered significant (equivalent to $p < 0.05$), after a process of adjustment for multiple testing

within distance d of the centre of each raster cell. Since we shall want to include the cell itself in the calculation, the weight w_{ii} will also be set to 1. The results of the G★ calculation can be expressed as z-values, which they are in Figure 11.13, and from which a p-value can be obtained by reference to the standard normal curve. As with the local Moran scores, the issue of repeat testing should be considered.[4]

Also of relevance is that the results are sensitive to the distance threshold, d. More generally, the problem is the results will change with the specification of the spatial weights matrix, which is true of any local or global statistic that incorporates one. In some cases there may be a theoretical justification for how the weights have been specified, or a review of the literature and of other comparable studies may be suggestive. In others, an optimisation procedure might be used. For example, if x_i is the number of homicides in grid cell i, and $x_{lag(i)}$ is the average number of homicides in the grid cells within distance, d, of I then we might look for a distance that minimises $\sum_{i=1}^{n} \left(x_i - x_{lag(i)} \right)^2$. This appears to be at about 1.5 km (see Figure 11.14).

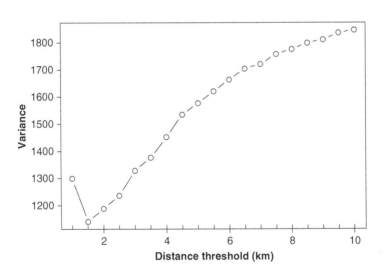

FIGURE 11.14   A variogram, which can be used to help decide which is the best distance threshold to use with the homicide data and with the local statistics

---

[4]Localised G★ statistics have been used to show that the enclaves occupied by minority ethnic groups have grown during the 10-year period between censuses in the UK, implying that the groups are not rooted to particular residences but are migrating outwards with a willingness to live amongst other ethnic groups and creating more mixed neighbourhoods (Johnston et al., 2015).

11.5 GEOGRAPHICALLY WEIGHTED STATISTICS

A local statistic simpler than either local Moran's I or $G\star$ and discussed in Section 5.3 is the average calculated for a location and its surrounding neighbours:

$$\bar{x}_{(u_i,v_i)} = \frac{\sum_{j=1}^{n} w_{ij} x_j}{\sum_{j=1}^{n} w_{ij}} \qquad (11.5)$$

where $w_{ij} = 1$ if locations i and j are considered to be neighbours, else $w_{ij} = 0$. In order for its own value to be included in the averaging, a location is considered to be a neighbour of itself ($w_{ii} = 1$).

Figure 11.15 shows the neighbourhood house prices for the Birmingham study region and also, at the top right, the local mean averages for when the centroids of the neighbourhoods have to be within 2 km of each other to qualify as neighbours. Defined this way, the average neighbourhood has five or six neighbours. However, there are some with fewer or no neighbours, as the connectivity graph at the top left of Figure 11.16 shows. The lack of connections could be resolved by increasing the distance threshold, but this does not address the underlying issue, which is the variable size of the Census neighbourhoods reflecting areas of greater or lesser population density. Increasing the distance threshold does not change the fact that some places are closer together than others and will have more neighbours. An alternative approach is to consider the k nearest other places as neighbours, regardless of how close or far away those places may be. This is sometimes referred to as an adaptive approach, because it adapts to the local density of the neighbourhoods – the search window will be smaller where the locations are close together and bigger where they are further apart. The local means shown in the middle left of Figure 11.15 use this approach for the six nearest neighbours (i.e. $k = 6$).

Even so, the weighting is an all-or-nothing approach: places are either neighbours ($w_{ij} = 1$) or they aren't ($w_{ij} = 0$). This could be modified on the basis that places that are closest together are more likely to exhibit similarities than those that are further apart, therefore giving most weight to the closest neighbours, lower weight to the further neighbours and zero weight to those that are either beyond the threshold distance or are not one of the k nearest neighbours. At its simplest, we can have:

$$w_{ij} = \begin{cases} 1/d_{ij} & \text{if } d_{ij} \leq d_{max} \\ 0 & \text{otherwise} \end{cases} \qquad (11.6)$$

where d_{max} is the threshold distance or the distance to the kth nearest neighbour of location i. This is an example of inverse distance weighting, where the greater

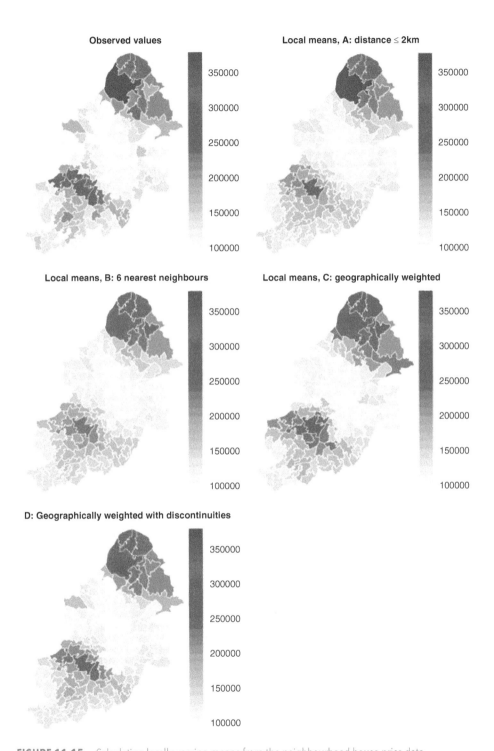

FIGURE 11.15 Calculating locally varying means from the neighbourhood house price data

Local means, A: distance ≤ 2km

Local means, B: 6 nearest neighbours

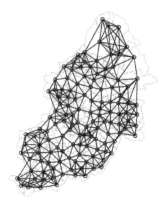

Local means, C: geographically weighted

D: Geographically weighted with discontinuities

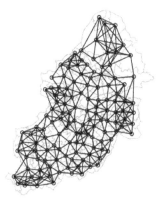

FIGURE 11.16 A connectivity graph showing which places are treated as neighbours in the calculation of the locally varying means

the distance between locations i and j, the lower the weighting. As it stands, the weights will tend towards but not reach zero for all locations within the distance, d_{max}. This can be altered so that they reach zero at d_{max}:

$$w_{ij} = \begin{cases} 1-(d_{ij}/d_{max}) & \text{if } d_{ij} \leq d_{max} \\ 0 & \text{otherwise} \end{cases} \tag{11.7}$$

This also avoids the potential division by zero that would occur in equation (11.6) when $d_{ij} = 0$ and the calculation error it would cause.

Both equations (11.6) and (11.7) create weights with a linear decay around location i, as shown in the top left of Figure 11.17. Alternative weightings have different shapes. For the bi-square weighting,

$$w_{ij} = \begin{cases} \left(1 - \left(d_{ij}/d_{max}\right)^2\right)^2 & \text{if } d_{ij} \leq d_{max} \\ 0 & \text{otherwise} \end{cases}$$

(11.8)

Other weightings and their formulae are given in Gollini et al. (2015).

The geographically weighted means shown in the middle right of Figure 11.15 were calculated using a bi-square weighting to the seventh nearest neighbour. Given that the weight is zero at the seventh neighbour, this means we are really using six (i.e. $k - 1$) nearest neighbours in the calculations.

In comparing with the observed values in Figure 11.15, it can be seen that the local means act to smooth out the geographical variations in the data. The greater the distance threshold or the greater the number of nearest neighbours, the greater the smoothing will be. The inverse distance weighting is motivated by Tobler's first

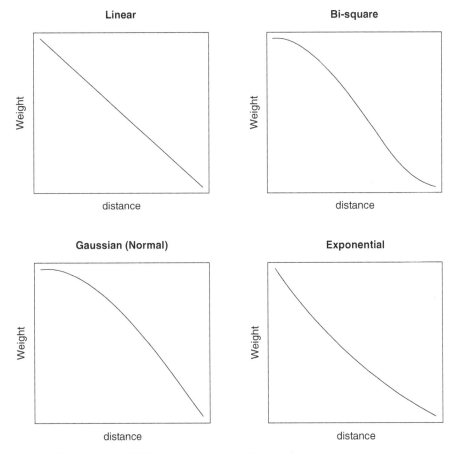

FIGURE 11.17   Examples of different types of inverse distance weighting

law of geography (see Chapter 10) and the belief that near things are more related than far things (positive autocorrelation). That is generally true for the neighbourhood average house prices, but there are also parts of the maps that show sharper differences between neighbouring locations (negative autocorrelation); see Mitchell and Lee (2014) and Harris (2014). In fact, it is not unusual to find parts of cities that are different from other areas around them – think of the Tenderloin in San Francisco, for example. The differences can be caused by pockets of social housing, the division of communities by roads or railways, ghettoisation, gentrification, the historical legacy of good or bad neighbourhoods, and so forth. With this in mind, we could say that near things are more related than far things, except for where the context means they aren't.

The bottom left of Figure 11.15 therefore modifies the geographically weighting to retain some of the spatial discontinuities observed in the data:

$$w_{ij} = \begin{cases} \left(1-\left(d_{ij}/d_{max}\right)^2\right)^2 & \text{if } d_{ij} \leq d_{max} \text{ and } \left|x_i - x_j\right| < z \\ 0 & \text{otherwise} \end{cases} \tag{11.9}$$

where $\left|x_i - x_j\right|$ is the difference in the neighbourhood average house price observed at i and j, and z is a maximum permitted difference.[5] The effect of allowing for the discontinuities can be seen by comparing local means C with local means D in the maps of Figure 11.15 and also the connectivity graphs of Figure 11.16.

You may be wondering about the purpose of calculating geographically weighted statistics given that we have the map of the original data anyway. There are three answers. First, in the presence of noisy data, the spatial smoothing can help to draw out the underlying geographical patterns in the data. Second, there is no necessity to limit the statistic to the geographically weighted mean (Brunsdon et al., 2002). The geographically weighted standard deviation or interquartile range, for example, will reveal localised variations in house prices, helping to detect some of the spatial discontinuities discussed above. Geographically weighted correlation and geographically weighted regression (see below) can explore geographical variations in the predictors of house prices. Third, the geographically weighted mean can be used as the basis for interpolation – that is, for predicting the average house price at sub-units of the original map. The problem with Census neighbourhoods as a unit of analysis is that they are discrete, vector objects that imply all change is at their boundaries, where the reality is more likely to be one of continuously varying house prices across the map. Figure 11.18 overlays a raster grid with cell length of

---

[5]For Figure 11.15 this was defined by splitting the observed values into quartile groupings and assigning a zero weight if neighbours were more than two quartiles apart.

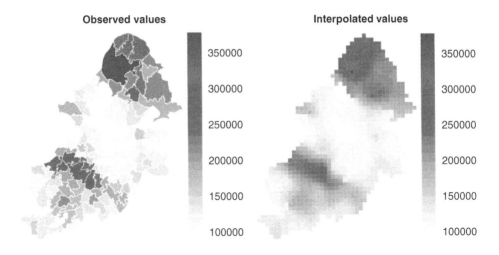

FIGURE 11.18   Using geographically weighted averaging for interpolation

500 metres over the original values and calculates the geographically weighted average in and around each cell, allowing for spatial discontinuities. The result moves from an object-based to more of a field-based view of how the house prices vary across the study region.

11.6 GEOGRAPHICALLY WEIGHTED REGRESSION

A natural extension to geographically weighted statistics is to ask whether the predictors of a measured Y variable have greater impact in some places than in others or everywhere are equal in their effect. This is the spirit of geographically weighted regression (GWR), which is an extension of the traditional regression framework discussed in Chapters 10 and 11. GWR works by comparing the results of a global regression model (fitted to the whole study region) to the set of results that are obtained if a local regression model is fitted at each individual location in turn (Fotheringham et al., 2002). The local regression model employs inverse distance weighting, such as the bi-square method of equation (11.8), giving most weight to the closest observations and zero weight to those that are either beyond a distance threshold or not one of the k nearest neighbours if an adaptive method is used. If there are n observations, then there are usually n different regression models to compare with each other, although the locations at which the models are fitted could be placed anywhere in the study region and need not be restricted to the n. In any case, the interest is in whether the estimated beta values emerging from the local regression analyses are approximately the same everywhere, meaning the effects of the X variables are broadly constant on the Y, or whether the relationships are spatially non-stationary, meaning they and vary in different parts of the map. Recall that in matrix form the formula for estimating the beta coefficients for ordinary least squares regression is given by equation (9.14),

$$\hat{\beta} = \left[\mathbf{X}^T\mathbf{X} \right]^{-1} \mathbf{X}^T\mathbf{y}$$

The equivalent for GWR is

$$\hat{\beta}_{(u_i,v_i)} = \left[\mathbf{X}^T\mathbf{W}_i\mathbf{X} \right]^{-1} \mathbf{X}^T\mathbf{W}_i\mathbf{y} \qquad (11.10)$$

where $\hat{\beta}_{(u_i,v_i)}$ is the estimated beta values for a subspace of the study region centred at location i and $\mathbf{W}_i$ is a spatial weights matrix specifying the weights attached to the other observations around i.

Figure 11.19 transports us away from the West Midlands of England, to the capital of China. The data are simulated but realistic and give, on a log scale, the land price at each of the locations shown (log of renminbi per square metre). The left-hand plot shows the locations of the land parcels but suffers from overplotting. These are the data that are used in the analysis that follows. The right-hand plot simplifies the data a little and is for visualisation only.

Within the simulated data set are some predictors of land value, including the size of the land parcel, distance to the central business district (CBD), distance to an elementary school, distance to a river, distance to a park and information about when the land parcel was sold. With the exception of the parcel size, all these variables are statistically significant at the 95 per cent confidence level. Focusing

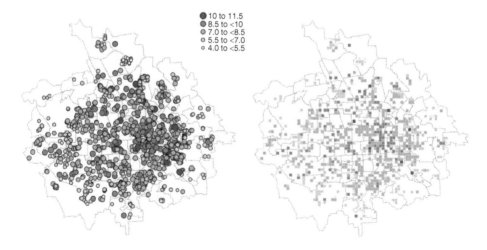

FIGURE 11.19 Simulated land price data for Beijing (log of renminbi per square metre)

on the distance to elementary school variable for the purpose of discussion, a one-kilometre increase in that distance entails an expected decrease in the land value of approximately 6 per cent.[6] The geographically motivated question is whether that expectation holds true across the study region or whether the average conceals important geographical variation.

To answer the question we turn to GWR. The process of GWR will typically begin by looking for a broadly optimal threshold distance or choice of k nearest neighbours for the inverse distance weightings. This is known as the bandwidth. Suppose that y_i is the actual value of Y observed at location i, and that $\hat{y}_i$ is the predicted value from a GWR that, for the purpose of calibration, is based only on the neighbours of i and not i itself. A preferred bandwidth could be the one that minimises $\Sigma_{i=1}^{n}|y_i - \hat{y}_i|$; the one for which the value of y_i is best predicted by the attributes of its neighbours.[7]

Having determined the bandwidth – here it is to the 286th neighbour of 1116 – the local and geographically weighted regression can be fitted in and around each location in turn. To reiterate, if there are n locations, then there are n estimates of the effect on land price of distance to an elementary school.[8] Figure 11.20 shows how the effect of distance to an elementary school varies

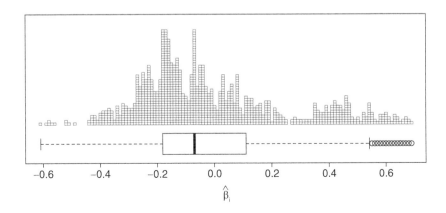

FIGURE 11.20 The distribution of the local beta values estimated by GWR for the distance to an elementary school variable

[6]Recall that when Y is measured on a log scale, the beta values can be interpreted as the percentage change in Y for a one-unit increase in the X (see Chapter 10).
[7]Other approaches include looking at the AIC scores of the models at different distance thresholds.
[8]As well as n estimates of all the other predictor variables unless it is chosen to hold one or more of these constant at their average for the study region (to not permit them to vary geographically).

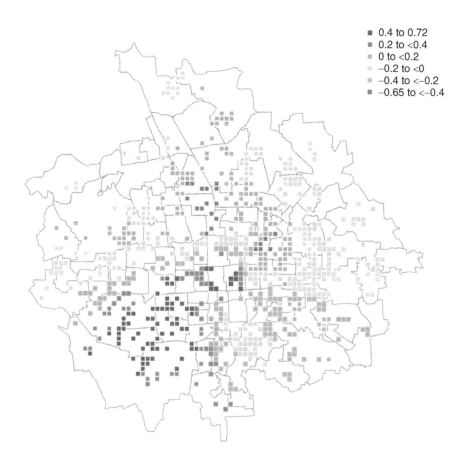

- ■ 0.4 to 0.72
- ■ 0.2 to <0.4
- ▦ 0 to <0.2
- ▨ −0.2 to <0
- ▨ −0.4 to <−0.2
- ■ −0.65 to <−0.4

FIGURE 11.21   Map of the local beta values for the distance to an elementary school variable. Those with a yellow border may be regarded as significant ($p < 0.05$)

across the study region, although it doesn't show where it varies. Figure 11.21 maps the estimates and shows there is a strong geographical patterning. In the south and especially southwest of the city, an increase in the distance to an elementary school is associated with an increased land value. In the north of the city it is associated with a decrease. A Monte Carlo procedure confirms that the spatial variation in the beta values is greater than that found under randomisation (by randomly moving the data around the locations in the study region).

Generally, accessibility to a school has a positive impact on land or house prices, so why should increased distance from an elementary school be associated with increased and not decreased prices for some parts of the city? A possible explanation is that there is something unmeasured about these places that makes them contextually different from the rest. It may be that although these places are further from the elementary schools, it is not this characteristic that actually raises the land value but some other variable that has not been included in the model. If that were the case, then the geographical patterning that is found in this as well as in the other variables has arisen because the model is misspecified – there are missing variables or perhaps some of the X variables are non-linear with the Y. The GWR would then serve as a useful diagnostic check on the standard regression model as well as providing an exploratory tool to reveal spaces of difference. However, it is also possible that in some locations distance from a primary school is indeed valued positively for whatever reason (perhaps the purchasers of the land are keen to keep away from children!). Less flippantly, there is little reason to suppose that even the most theoretically informed model should reveal effect sizes that are constant everywhere. For example, the interaction of people within neighbourhoods can generate peer or neighbourhood effects that alter the relationship between a Y and an X variable in regionally specific ways.

11.7 MULTILEVEL MODELLING

GWR adopts a field-based view of spatially varying relationships, allowing them to change continuously across the study region. An alternative approach is to treat the observations as nested within the boundaries of discrete objects at a higher geographical scale and to look at the variations between those objects. For the (simulated) land parcel data, the parcels are situated within Census districts, shown with a grey boundary in Figures 11.19 and 11.21. What we have is a two-level geographical hierarchy, with the land parcels at level 1 and the districts at level 2.

One way to model the differences at the higher level is to fit what is known as a fixed effects model of the form

$$y = \hat{\beta}X + \hat{\delta}Z + \hat{\varepsilon} \qquad\qquad (11.11)$$

where the Zs are a series of binary (dummy) variables, one for each of the districts, an observation being given the value 1 if the land parcel falls within the district, otherwise 0. For example, if $z_{11} = 1$ it means that land parcel 1 is in district 1, whereas $z_{21} = 0$ means land parcel 2 isn't; if land parcel 2 is in district 3 then $z_{23} = 1$. Fitting the model allows us to see if the average price of a land parcel is higher in some districts than in others, having allowed for the effects of the X Variables. The expected increase or decrease in district 1 is the estimated value, $\hat{\delta}_1$, in district 2 it is $\hat{\delta}_2$, and so forth. What we are doing is fitting a separate Y-intercept for each of the higher-level districts. For district 1 the intercept is at $\hat{\beta}_0 + \hat{\delta}_1$, for district 2 it is at $\hat{\beta}_0 + \hat{\delta}_2$, etc. The left-hand plot of Figure 11.22 maps the differences and reveals a spatial pattern which could be investigated further using Moran's I ($I = 0.188$, $p < 0.001$), local Moran's I and other measures of spatial dependency.

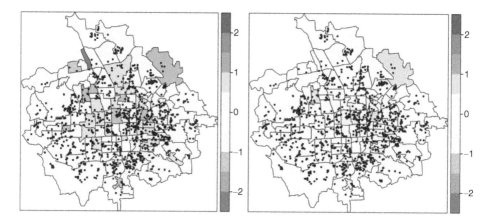

FIGURE 11.22 Estimates of the district-level variations in land parcel prices, having controlled for the X variables (left). From a fixed effects model, where the differences between places are measured using dummy variables in a standard regression. From a random effects model, using a multilevel model to measure variances at the district level of the hierarchy (right)

In principle the model could be extended to also allow the slopes of one or more of the X variables to vary from district to district.[9] However, there are limitations in doing so. Firstly, some districts contain one or a few observations, so the estimation is either impossible or highly uncertain. Second, the model would lack parsimony. There are 111 districts in the study region. That means 111 dummy variables to estimate the intercepts and a further 111 variables per X variable if the slope is allowed to vary too.

Multilevel modelling tackles the issue in a different way. Consider the model

$$y_{ij} = \hat{\beta}_0 + \hat{\beta}_1 x_{1ij} + \hat{\beta}_2 x_{2ij} + \cdots + \hat{\varepsilon}_{ij}$$

$$(11.12)$$

where the subscripts i and j denote a two-level hierarchy. The residual errors can be split into variation around the intercept and between the districts, and the variation that remains within the districts, at the land parcel level, after the district-level variation has been taken into account:

$$y_{ij} = \hat{\beta}_{0j} + \hat{\beta}_1 x_{1ij} + \hat{\beta}_2 x_{2ij} + \cdots + \hat{\varepsilon}_{ij}$$

$$\hat{\beta}_{0j} = \hat{\beta}_0 + \hat{\mu}_{0j}$$

$$(11.13)$$

---

[9]This would be achieved by creating interaction terms: by multiplying the X variables by the dummy variables and including those interactions in the model.

The higher-level residuals (μ_j) are considered to be random – that is, as a set of outcomes from a distribution of possible values that usually is assumed to be normal.[10] It is known as a random intercepts model. An advantage of multilevel modelling is that it allows the variation in the residuals to be estimated and compared at each level of the hierarchy net of the other levels, allowing the within and between area variations to be considered simultaneously.

Having fitted a model with the same predictors of land price as before, 21 per cent of the remaining residual variation is found to be between districts – a reasonable amount. A map of the district-level variation is on the right of Figure 11.22. Comparing that map to the one for the fixed effects model, the spatial pattern is broadly the same but some of the extremes have been curtailed – helpfully so where they were based on only one or two observations at the land parcel level. The district-level residuals can also be shown as a caterpillar plot with the 95 per cent confidence interval drawn around each estimate (Figure 11.23). We may conclude that there are significant between-district differences in the land prices that the predictor X variables have yet to account for. The added bean plot suggests the assumption of normality was credible.

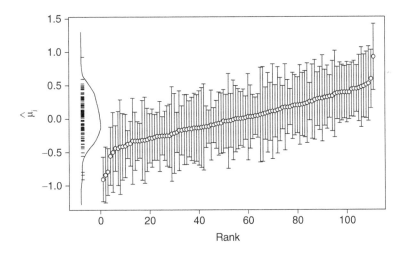

FIGURE 11.23 A caterpillar plot of the district-level residuals

Multilevel models can also allow the slope of any of the X variables to vary from place to place in the hierarchy. For example,

---

[10]The same is true of the lower-level residuals.

$$y_{ij} = \hat{\beta}_{0j} + \hat{\beta}_{1j} x_{1ij} + \hat{\beta}_2 x_{2ij} + \cdots + \hat{\varepsilon}_{ij}$$

$$\hat{\beta}_{0j} = \hat{\beta}_0 + \hat{\mu}_{0j}$$

$$\hat{\beta}_{1j} = \hat{\beta}_1 + \hat{\mu}_{1j} \tag{11.14}$$

which is a random intercepts model with a random slope also. Applying this to the distance to elementary school variable, the districts where it has a positive relationship with land prices and the districts where it has a negative relationship are shown in red and blue respectively on the left in Figure 11.24. On the right the GWR estimates are redrawn for ease of comparison. They are broadly similar but not identical. That they differ is not surprising: they are different approaches rooted in conceptually different views of how the geographical processes are structured and contained (an object/discrete-based view versus a field/continuous one).

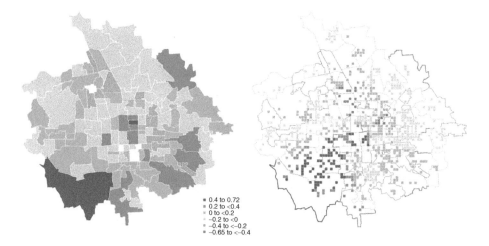

0.4 to 0.72
0.2 to <0.4
0 to <0.2
−0.2 to <0
−0.4 to <−0.2
−0.65 to <−0.4

FIGURE 11.24   The district-level variations in the effect of distance to an elementary school on land prices, estimated as a random slope model (left). The GWR estimates (right)

Multilevel modelling offers a powerful and flexible framework for analysis that has proved very popular in geography where data often have hierarchical structures – people nested into neighbourhoods nested into cities nested into regions, for example, or surveys that are undertaken using a clustered sampling method. It offers a number of practical benefits.[11]

---

[11]See Owen et al. (2015), for a critical review.

First, the number of levels is not limited to two but can be extended to incorporate whatever are relevant to the data and to the analysis. This allows investigation of the scale at which a geographical process operates. For example, segregation can be measured as variation within and between places and multilevel modelling can be used to assess whether segregation is most prevalent at the neighbourhood, city or regional scale (Leckie and Goldstein, 2014; Jones et al., 2015).

Second, multilevel modelling allows variables that have been measured at different scales of aggregation to be introduced into the model at the appropriate level of the hierarchy. Imagine, for example, that we have knowledge of the population density per Census district and this was pertinent to the price of land parcels in the simulated data for Beijing. The population density is a level 2 variable. This is a problem for classic regression modelling because it necessarily introduces a group dependency: for any district, the land parcels that fall within it will be tagged with the same population density. Multilevel modelling allows for these group dependencies to be handled properly.

Third, multilevel modelling is not limited to modelling geographical hierarchies. The hierarchy could be temporal or it could be a mix of the temporal and the spatial. This is especially useful for analysing longitudinal data, with repeated measurements of people or places over time.

Fourth, multilevel models can model complex hierarchies, for example those that overlap geographically or where a observation at one level is a member of multiple groups at another.

Fifth, multilevel models can handle incomplete data, when some measurements are missing.

Nevertheless, all methods have limitations. Multilevel modelling is not a fully geographical method of analysis, in the sense that it does not consider the geographical coordinates of where the measurements were observed and the distances between the observations in the form of an explicit spatial weights matrix. The geographical relationships are implied by the hierarchical modelling structure, but moving the locations of the land parcels within any one district would make no difference to the results of the multilevel model, whereas it would to the GWR.

Second, multilevel modelling is sometimes used to search for neighbourhood effects, a rather loose phrase that suggests that where people live has an impact upon their choices, opportunities, health, educational outcomes and so forth. This could be conceived in terms of a first-order and functional relationship where, most simply, X leads to Y, so if X happens to be concentrated in particular places then we can expect Y to be so too. However, neighbourhood effects actually imply something more than this: a second-order effect where there is something about the interaction between people living in close proximity to each other that raises or lessens the impact of X upon Y. A weakness of multilevel modelling is that is can be used to highlight geographical differences but is not really a process-based model of how those differences arose. The same is true of GWR but not of the spatially lagged Y model of Chapter 10, which is explicitly positing a process of overspill from one area to its neighbours. Multilevel models consider 'vertical' (group or contextual) effects but not 'horizontal' (spatial interaction) effects.

In recent years, there has been some coming together of multilevel and spatial econometric approaches. For example, Dong and Harris (2015) develop the catchy-sounding hierarchical spatial autoregressive model, which allows for the separate estimation of horizontal effects between places from the vertical effects acting upon places, thereby preventing them from being confused and conflated. If we denote a standard two-level random intercepts model as

$$\mathbf{y} = \mathbf{X}\boldsymbol{\beta} + \boldsymbol{\mu} + \boldsymbol{\varepsilon} \tag{11.15}$$

where $\boldsymbol{\mu}$ and $\boldsymbol{\varepsilon}$ are the level 1 and 2 residuals respectively, and a standard spatially lagged Y model as

$$\mathbf{y} = \rho\mathbf{W}\mathbf{y} + \mathbf{X}\boldsymbol{\beta} + \boldsymbol{\varepsilon} \tag{11.16}$$

where $\mathbf{W}$ is the spatial weights matrix, then the two approaches may be combined to give

$$\mathbf{y} = \rho\mathbf{W}\mathbf{y} + \mathbf{X}\boldsymbol{\beta} + \boldsymbol{\mu} + \boldsymbol{\varepsilon} \tag{11.17}$$

or, more generally,

$$\mathbf{y} = \rho\mathbf{W}\mathbf{y} + \mathbf{X}\boldsymbol{\beta} + \gamma\mathbf{Z} + \theta + \boldsymbol{\varepsilon}$$

$$\theta = \zeta\mathbf{M}\mathbf{y} + \boldsymbol{\mu} \tag{11.18}$$

where $\mathbf{X}$ and $\mathbf{Z}$ are the level 1 and level 2 predictors, respectively ($\boldsymbol{\beta}$ and γ their associated coefficients), $\boldsymbol{\mu}$ and $\boldsymbol{\varepsilon}$ are as before, and ρ and ζ are measures of spatial autocorrelation (spatial interaction) at levels 1 and 2, with spatial weights $\mathbf{W}$ and $\mathbf{M}$. The method has been used to study spatial interaction effects in the emerging land market of Beijing (Dong et al., 2015) and also social interaction effects in food expenditures in Poland (Laszkiewicz et al., 2014).

11.8 CONCLUSION

This chapter has looked at geographical methods of analysis, characterised by their consideration to where the data have been collected as well as to the measurements themselves. It has discussed various methods of analysis, some of which are concerned with finding patterns of spatial clustering (such as the point pattern analyses) and some with revealing spatial heterogeneity (such as GWR, which identifies spatial variation in regression relationships). Some methods employ a continuous and field-based view of geographical space (GWR again) whereas others default to a more object-based view (multilevel modelling). Some might be regarded as global methods with a single, region-wide measure of spatial association (the spatial autocorrelation term in the spatial econometric models of Chapter 10, for example, or the standard Moran's value) whereas others look for localised patterns of spatial dependence (LISAs and localised Moran's). Most have a spatial weights matrix central to their calculation but for multilevel modelling the geography is implied by the geographical hierarchy of the objects being analysed. What they all have in common is that the locational aspect of geographic data is intrinsic to and not discarded in their methods of analysis. That is what makes them geographical.

These sorts of analyses are not necessarily standard to some of the better-known statistical packages that are sometimes used in university teaching. Other software such as ArcGIS (Spatial Statistics Toolbox) or SpaceStat do offer it, but at a price. However, there are also various free software packages, including GeoDa, PySAL and GeoDaSpace, available from the GeoDa Center for Geospatial Analysis and Computation,[12] and MLwiN for multilevel modelling (unfortunately the latter is only free within the UK).[13] Usefully, the open source software, R, offers a range of add-in libraries that extend its base functionality to enable mapping, point pattern analysis, geostatistics, geographically weighted statistics, spatial regression, GWR, and other geographically minded approaches. A brief overview of R is presented in the next chapter.

---

[12]https://geodacenter.asu.edu/software
[13]http://www.bristol.ac.uk/cmm/software/mlwin/

12

AN INTRODUCTION TO R

12.1 INTRODUCTION

Figure 1.2, in Chapter 1, introduced seven components of quantitative geography: social and scientific knowledge; maths and numeracy; statistics and statistical modelling; geographical information science; visualisation and data presentation; thinking geographically and looking at spatial variation; and data handling and (geo)computation. Of these, computation has taken a back seat. It has been there, of course, in the analyses, the calculations and the production of the graphics (you will have realised that I did not do them by hand). However, there has been no specific discussion of why computational skills and programming are important skills to learn and to apply.

You might dispute that they are. After all, the trend in portable devices has been to make the interface as graphical and as user-friendly as possible – these days it's not even point-and-click but touch-and-swipe. You don't need to write computer code to use your smartphone, so why should you need to type lines of syntax in order to undertake quantitative analysis?

Possibly you don't. Plenty of statistical, modelling and GIS software packages offer drop-down menus that you can select without any particular need for programming. However, there are three problems with reliance on this sort of approach. The first is that few data sets come ready to use without the need for some sort of checking and cleaning of the data. That tidying might require a little bit of coding to manipulate the data. Second, what if your analysis is not entirely routine? What if you want to take it in directions that the menu interface has not considered and does not easily allow for? What if you want more direct control over the process or to customise things your way? Third, what if you or others want to reproduce the analysis or make some changes to it? The advantages of moving a mouse to select drop-down menus wear thin if you have to do it over and over again, requiring you to remember exactly which buttons need to be pressed in order to get to the output. Far easier to have the process written down as a script that can be rerun with or without alternations. It is also more open and scientific in the sense that a guiding principle behind science is that the work should be reproducible so that others can check it, build upon it, debate it and look for different conclusions (see Brunsdon, 2015). That is especially important with big and complex data sets that are sifted, sorted and modified in ways that should be made clear to others and not just hidden behind the curtains of data processing.

This chapter focuses on R as a programming language that is relevant to geography and worth taking time to learn. Of course, it's not the only programming language but it does have a number of advantages that are discussed below. The aim of this chapter is not to provide a comprehensive introduction to R, but to provide a flavour of what it is like, what it can do, and why it is a good tool for geographical analysis. Follow-up readings are suggested in the conclusion to help extend your knowledge.

12.2 WHY R?

R is described on its website as 'a free software environment for statistical computing and graphics. It compiles and runs on a wide variety of UNIX platforms, Windows and MacOS.'[1] R has developed over many years, is widely used (by businesses as well as by researchers) and has extensive documentation, including well-written help files with example code and applications, and a large user community. There are many reasons to be interested in R. Being free is an advantage, but better is its open source nature. That allows you to see the code behind its functions, to adapt it, to borrow other people's code and to share your own. The open nature encourages people to write libraries that greatly extend R's base functionality. The libraries used in this book and the purpose for which they were used are listed in Table 12.1. The functionality that some of these offer for geographical analyses and for GIS-style handling of geographical data has helped swell the popularity of R within quantitative geography.

A disadvantage of R is that it has a learning curve: it takes a while to get used to, and not everything is immediately obvious. There are often different ways of achieving the same objective, which provides flexibility but can be confusing. The commands are not always consistent from one library to the next, although there is increasingly quite a high level of standardisation.

Installing R

R may be obtained from the Comprehensive R Archive Network, CRAN.[2] Once installed, it can be run using its own graphical user interface or in conjunction

---

[1]https://www.r-project.org/
[2]https://cran.r-project.org/

TABLE 12.1   The additional R libraries used in the production of this book

Library name	Application in this book
beanplot	Drawing bean plots
classInt	Creating map classes (splitting the data into equal intervals, quantiles, etc.)
data.table	Reading in large data sets quickly
fmsb	Creating a radar chart
fMultivar	Creating plots with hexagonal binning
foreign	Reading in data from other statistical packages
gdistance	Least-cost route calculation
GISTools	Creating choropleth maps
gridExtra	Drawing some types of map next to each other on the page
GWmodel	Calculating geographically weighted statistics
lme4	Multilevel modelling
lmtest	Diagnostic tests for linear regression models
maptools	Reading and handling shapefiles
MASS	Robust regression
nlme	Multilevel modelling
nortest	Tests for normality
osmar	To access Open Street map data
plotrix	'3D' pie chart
pyramid	Pyramid plots
quantreg	Quantile regression
raster	Reading, creating and manipulating raster data
RColorBrewer	Creating colour palettes for maps and other graphics
rgdal	Changing coordinate reference systems
rgeos	Vector data analysis (geometric operations)
RgoogleMaps	Downloading and displaying Google map tiles
scatterplot3d	Creating a 3D scatter plot
sp	Handling and plotting spatial data
spatstat	Spatial point pattern analysis
spdep	Creating spatial weights matrices and spatial regression
twitteR	Obtaining Twitter data
vioplot	Drawing violin plots
wordcloud	Drawing word clouds

with RStudio, an integrated development environment for R.[3] R is command-line software: a command is typed and executed by the software. It is helpful to keep a log of those commands – to write and to save them as a script that can be rerun in R without having to type the commands again.

Within R, additional libraries can be installed using the `install.packages()` function. For example,

```
install.packages("raster")
```

will install the raster handling package. Most of the libraries of relevance to spatial analysis are listed on the spatial task view page.[4] A shortcut way of installing these packages is:

```
install.packages("ctv")
require(ctv)
install.views("spatial")
```

Installing packages requires an internet connection. Packages only need to be installed once, but should be activated each time R is restarted and the package is required. For example,

```
require(raster)
# Makes the raster package available for use
```

12.3 THE BASICS OF R

At its simplest, R can be used as a calculator. Typing 1 + 1 after the prompt > will produce the result 2 (after pressing the Return key, ↵), as in the following example:

```
1 + 1
[1] 2
```

Comments are indicated with a hash tag and will be ignored:

```
# This is a comment, no need to type it
```

Some other simple mathematical expressions are given below:

---

[3]https://www.rstudio.com/products/RStudio/
[4]https://cran.r-project.org/web/views/Spatial.html

```
10 - 5
[1] 5
10 * 2
[1] 20
10 - 5 * 2
[1] 0
# The order of operations gives priority to multiplication, as it
# should
(10 - 5) * 2
[1] 10
# The use of brackets changes the order
sqrt(100)
[1] 10
# Calculates the square root
10^2
[1] 100
# i.e. 10 squared
100^0.5
[1] 10
# Another way of calculating the square root
10^3
[1] 1000
log10(100)
[1] 2
# Calculates the log of 100 with a base of 10
log10(1000)
[1] 3
100 / 5
[1] 20
100^0.5 / 5
[1] 2
```

Incomplete commands

If you see the + symbol instead of the usual > prompt, what has been typed is incomplete. Often there is a missing bracket. For example,

```
sqrt(
+ 100
+ )
[1] 10
# The + symbol indicated that the command was incomplete
(1 + 2) * (5 - 1
+ )
[1] 12
```

Commands broken over multiple lines can be easier to read:

```
for (i in 1:10) {
+ print(i)
+ }
[1] 1
[1] 2
[1] 3
[1] 4
...

# A simple loop printing the numbers 1 to 10 on screen
```

Repeating or modifying a previous command

If a mistake needs to be corrected or if some previously typed commands are to be repeated then the ↑ and ↓ keys on the keyboard may be used to scroll between previous entries in the R Console.

Scripting

You can create a new script file from the drop-down menu using File → New script (in Windows) or File → New Document (Mac OS). It is a text file in which you can write a sequence of commands. For example,

```
a <- 1:10
print(a)
```

In Windows, if you highlight the commands and press Ctrl + R, they will be run in the R Console. If you continue to keep the focus on the Scripting window and go to Edit in the RGui you will find an option to run everything. Similar commands are available for other operating systems (e.g. Mac key + Return). You can save files and load previously saved files. Scripting is both good practice and good sense. It is good practice because it allows for reproducibility of your work. It is good sense because if you need to go back and change things you can do so easily without having to start from scratch.

Logging

You can save the contents of the R Console window to a text file that will then provide a log file of the commands you have been using (including any mistakes). The easiest way to do this is to click on the R Console (to take the focus from the Scripting window) and then use File → Save History (in Windows) or File → Save As (Mac). Note that graphics are not usually plotted in the R Console and therefore need to be saved separately.

Functions, assignments and getting help

It is helpful to understand R as an object-oriented system that assigns information to objects within the current workspace. The workspace is simply all the objects that have been created or loaded since beginning the session. Look at it this way: the objects are like box files, containing useful information, and the workspace is a larger storage container, keeping the box files together. A useful feature of this is that R can operate on multiple tables of data at once: they are just stored as separate objects within the workspace.

To view the objects in the workspace, type

```
ls()
```

Doing this runs the function `ls()`, which lists the contents of the workspace. The result `character(0)` indicates that the workspace is empty, assuming it is. To find out more about a function, type `?` or `help` with the function name:

```
?ls()
help(ls)
```

Either will provide details about the function, including examples of its use. It will also list the arguments required to run the function, some of which may be optional and some of which may have default values that can be changed as required. Consider, for example,

```
?log()
```

A required argument is `x`, which is the data value or values. Typing `log()` omits any data and generates an error. However, `log(100)` works fine. The argument base takes a default value of *e*, which means the natural logarithm is calculated. The default is assumed unless otherwise stated, so `log(100)` is a shortened version of `log(100, base=exp(1))`, whereas `log(100, base=10)` gives the common logarithm. The latter can also be calculated using the convenience function `log10(100)`.

The results of mathematical expressions can be assigned to objects, as can the outcome of many commands executed in R. When the object is given a name different than other objects within the current workspace, a new object will be created. Where the name and object already exist, the previous contents of the object will be overwritten, without warning.

```
a <- 10 - 5
print(a)
[1] 5
b <- 10 * 2
print(b)
[1] 20
print(a * b)
[1] 100
a <- a * b
print(a)
[1] 100
```

In these examples the assignment is achieved using the combination of < and -, as in a `<-` `100`. Alternatively, `100` `->` `a` could be used or, more simply, a `=` `100`. The `print(...)` command can usually be omitted, though it is useful, and sometimes necessary.

```
f <- a * b
print(f)
[1] 2000
f
[1] 2000
sqrt(b)
[1] 4.472136
print(sqrt(b), digits=3)
[1] 4.47
# The additional parameter now specifies the number of significant
    figures
```

```
c(a,b)
[1] 100 20
# The c(…) function combines its arguments
c(a,sqrt(b))
[1] 100.000000 4.472136
print(c(a,sqrt(b)), digits=3)
[1] 100.00 4.47
```

Naming objects in the workspace

Although the naming of objects is flexible, there are some exceptions:

```
_a <- 10
Error: unexpected input in "_"
2a <- 10
Error: unexpected symbol in "2a"
```

Note also that R is case sensitive, so a and A are different objects:

```
a <- 10
A <- 20
a == A
[1] FALSE
# == is a logical query: is a equal to A?
```

Removing objects from the workspace

To remove an object from the workspace it can be referenced by name:

```
ls()
rm(A)
```

To delete all the objects in the workspace and to empty it, type the following code, but be warned that there is no undo function. Whenever rm() is used the objects are deleted permanently.

```
rm(list=ls())
ls()
character(0)
# In other words, the workspace is empty
```

Saving and loading workspaces

Because objects are deleted permanently, a sensible precaution is to save the work-space prior to using rm(). To do so permits the workspace to be reloaded if necessary and the objects recovered. One way to save the workspace is to use

```
save.image(file.choose(new=T))
```

Alternatively, the drop-down menus can be used (File → Save Workspace in the Windows version of the RGui). In either case, you may need to type the extension. (RData) manually to prevent it being omitted, making it harder to locate and reload what has been saved. To load a previously saved workspace, use

```
load(file.choose())
```

or the drop-down menus.

When quitting R, it will prompt to save the workspace image. If the option to save is chosen it will be saved to the file .RData within the working directory. Assuming that directory is the default one, the workspace and all the objects it contains will be reloaded automatically each and every time R is opened, which could be useful but also irritating. To stop it, locate and delete the file. The current working directory is identified using the get working directory function, getwd() and changed most easily using the drop-down menus but see also ?setwd().

```
getwd()
[1] "/Users/rich_harris"
```

(Your working directory will differ from the above)

Quitting R

Before quitting R, you may wish to save the workspace. To quit R use either the drop-down menus or

```
q()
```

As promised, you will be prompted whether to save the workspace. Answering yes will save the workspace to the file .RData in the current working directory. To exit without the prompt, use

```
q(save = "no")
```

Or, more simply,

```
q("no")
```

12.3 A GEOGRAPHICAL INTRODUCTION TO R

Having explored the fundamentals of R, this section provides a short introduction to some of its geographical capabilities. The idea is to showcase some of what R can do. Not everything is explained in full, but the comments provide guidance. Begin by installing R and the libraries classInt, GISTools, gridExtra, GWmodel, rgdal, RgoogleMaps and spdep.[5] Then download the files birmingham.shp, birmingham.shx and birmingham.dbf from http://bit.ly/1LjsJ6T. Together they form a shapefile of the Birmingham study region shown in Figure 10.1 and containing the neighbourhood average house prices. To read the map data into R:

```
require(GISTools)
mymap <- readShapePoly(file.choose())
```

and select birmingham.shp. Assuming you have installed the GISTools library, this will read the shapefile into an object in R's workspace called mymap. You can check it is there by typing ls() and can confirm that it is a polygonal, spatial object with attribute data attached to it by typing class(mymap).

Looking at the attribute data, it contains a variable, price_paid, which is the neighbourhood average house price:

```
head(data.frame(mymap))
```

If we wish to see how the variable is distributed then there are various ways we can do so. For example,

```
summary(mymap$price_paid)
# The use of the $ is to reference a specific variable contained
  # within the object mymap
hist(mymap$price_paid)
boxplot(mymap$price_paid, horizontal=T)
```

Note that with the boxplot function, one of its arguments is whether to draw it horizontally or not. The default is not but it can be changed, as in the example.

---

[5]I am using R version 3.1.3

Plotting the data

Since mymap is a spatial object, a simple choropleth map of the average house prices can be produced using

```
spplot(mymap, "price_paid")
```

If you don't like the default colours, you can go through the process of creating your own using ColorBrewer. For example:

```
require(classInt)
nclasses <- 6
# The number of map classes
brks <- classIntervals(mymap$price_paid, nclasses, style="equal")
# Equal interval map classes are chosen
brks <- brks$brks
# Extracts the break points for the map classes. There is always one
# more break than there are map classes
brks[nclasses + 1] <- brks[nclasses + 1] + 0.001
# Slightly increases the upper break to ensure all areas are
# shaded on the map
palette <- brewer.pal(6, "YlOrRd")
# Creates the colour palette. See ?brewer.pal
spplot(mymap, "price_paid", at=brks, col.regions=palette)
```

Overlaying the map on a Google map tile

A more ambitious undertaking is to display the choropleth map on top of a Google Map tile. First, we need to set the coordinate reference system for the map, using the EPSG code (see Chapter 8) for the British National Grid System:

```
proj4string(mymap) <- CRS("+init=epsg:27700")
```

Next the map is transformed into longitude and latitude:

```
require(rgdal)
mymap <- spTransform(mymap, CRS("+init=epsg:4326"))
```

The RgoogleMaps library is then used:

```
require(RgoogleMaps)
bb <- bbox(mymap)
# Finds the bounding box for the map
gmap <- GetMap.bbox(bb[1,], bb[2,])
# Downloads the map tile
map.class <- cut(mymap$price_paid, brks, labels=F, include.lowest=T)
# Splits the data into groups based on the equal interval
# classification created earlier
palette <- AddAlpha(palette)
# Adds transparency (an alpha value) to the colour palette
PlotOnStaticMap(gmap)
# Displays the map tile
for(i in 1: nclasses) {
polys <- mymap[map.class==i,]
polys <- as(polys, "SpatialPolygons")
PlotPolysOnStaticMap(gmap, polys, col=palette[i], add=T)
}
# Loops through the map classes and adds them to the map
# one group at a time
```

The result is shown in Figure 12.1.

Testing for spatial autocorrelation

The map suggests there is a spatial patterning to the average house prices. We could look at this using Moran's test. To do so, we first need to create a spatial weights matrix, based on contiguity of neighbourhoods, for example:

```
mymap <- spTransform(mymap, CRS("+init=epsg:27700"))
# Return the coordinates back to British National Grid
require(spdep)
contig <- poly2nb(mymap, snap=1)
# Create a list of contiguous areas
plot(mymap)
plot(contig, coordinates(mymap), add=T)
```

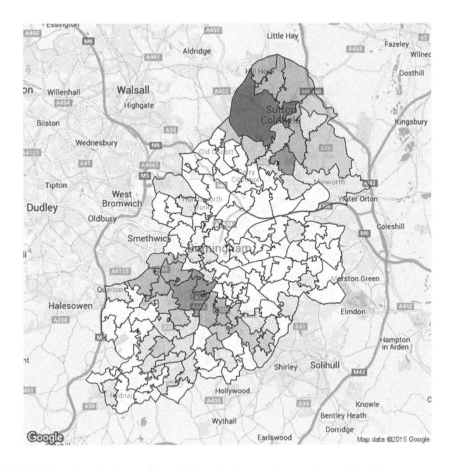

FIGURE 12.1   Choropleth map of the neighbourhood average house prices in Birmingham

```
# Draws the connectivity graph (as in Figure 11.16)
spweights <- nb2listw(contig)
# Converts the list of neighbours into row-standardised spatial weights
summary(spweights)
#  Provides  information  about  the  weights,  for  example  the
#  average number of neighbours
```

Once we have the spatial weights matrix we can carry out a Moran test and create a Moran plot of the data, either as it is or as the log of the average house price:

```
moran.test(mymap$price_paid, spweights)
moran.plot(mymap$price_paid, spweights)
```

```
moran.test(log(mymap$price_paid), spweights)
moran.plot(log(mymap$price_paid), spweights)
```

Local Moran values

We can also create a map of the local Moran *I* values:

```
localI <- localmoran(log(mymap$price_paid), spweights,
  p.adjust.method="holm")
# Calculates the local Moran values and also makes an adjustment
# to the p-values for multiple testing
head(localI)
# Takes a look at the results
palette <- auto.shading(localI[,1], cutter=rangeCuts, n=6,
  cols=brewer.pal(6, "YlOrRd"))
# Creates a colour palette
choropleth(mymap, localI[,1], shading=palette)
# Plots a choropleth map of the local Moran values
# which are in the first column of localI
choro.legend(395637, 300766, sh=palette)
# Adds a legend to the map
submap <- mymap[localI[,5] < 0.05,]
plot(submap, border="dark blue", lwd=3, add=T)
# Highlights the significant values
```

The map should look like Figure 12.2. The results are, of course, dependent upon the spatial weights matrix. Instead of contiguity it could, instead, be based on a *k* nearest neighbours approach:

```
knn <- knearneigh(coordinates(mymap), 6)
# Here the six nearest neighbours are searched for
knn <- knn2nb(knn)
spweights <- nb2listw(knn)
```

Geographically weighted statistics

Alternatively, we could calculate and map geographically weighted statistics, producing the results shown in Figure 12.3:

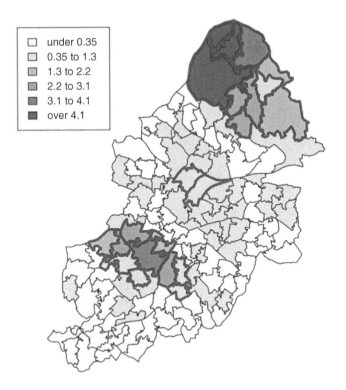

FIGURE 12.2 Local Moran values. Statistically significant values ($p < 0.05$, after adjustment for multiple testing) are shown with a blue border

```
require(GWmodel)
dmatrix <- gw.dist(coordinates(mymap))
# Calculates the distances between the centroids (centres) of each
# area
gwstats  <-  gwss(mymap,  vars="price_paid",  bw=7,  adaptive=T,
  dMat=dmatrix)
# Calculates the geographically weighted statistics
names(gwstats$SDF)
# shows the names of the variables in gwstats$SDF
require(gridExtra)
plot1 <- spplot(gwstats$SDF, "price_paid_LM", col.regions=color-
  RampPalette(c("yellow","orange","red"))(100))
plot2 <- spplot(gwstats$SDF, "price_paid_LSD", col.regions=color-
  RampPalette(c("yellow","orange","red"))(100))
grid.arrange(plot1, plot2, ncol=2)
```

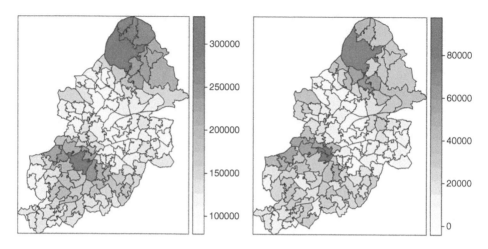

FIGURE 12.3 The geographically weighted mean (left) and standard deviation (right) of the house price data

Regression, spatial regression and geographically weighted regression

The relationship between the log of the neighbourhood average house price and various predictor variables can be considered by fitting an ordinary least squares regression model. Here the Xs are the percentage of the population unemployed and the proportion of properties sold that were detached:

```
ols1 <- lm(log(price_paid) ~ pctunemp + detached, data=data.
   frame(mymap))
# Fits the linear model
summary(ols1)
# Estimates of the beta coefficients, R-square, etc.
par(mfrow=c(2,2))
# Creates a 2 by 2 plotting window for the plots
plot(ols1)
# Diagnostic plots
par(mfrow=c(1,1))
# Resets the plot window
```

The residuals can be checked for evidence of spatial autocorrelation:

```
palette <- auto.shading(rstandard(ols1), cutter=rangeCuts, n=6,
   cols=brewer.pal(6, "PRGn"))
choropleth(mymap, rstandard(ols1), shading=palette)
choro.legend(395637, 300766, sh=palette)
# Map of the (standardised) residuals
contig <- poly2nb(mymap, snap=1)
spweights <- nb2listw(contig)
lm.morantest.exact(ols1, spweights)
# Tests the residuals for spatial dependency
```

Other models can then be tried:

```
sperr <- errorsarlm(log(price_paid) ~ pctunemp + detached,
   data=data.frame(mymap), spweights)
# Fits a spatial error model
lagY <- lagsarlm(log(price_paid) ~ pctunemp + detached,
   data=data.frame(mymap), spweights)
# Fits a spatially lagged Y model
summary(lagY)
summary(sperr)
AIC(ols1)
AIC(sperr)
AIC(lagY)
# AIC estimates of the model fits
impacts(lagY, listw=spweights)
# Impact measures for the spatially lagged Y model
bw <- bw.gwr(log(price_paid) ~ pctunemp + detached, data=mymap,
   adaptive=T)
gwr1 <- gwr.basic(log(price_paid) ~ pctunemp + detached, data=mymap,
   adaptive=T, bw=bw)
gwr1
# Geographically weighted regression
```

The spatially varying effects of the X variables can also be mapped, although not all are necessarily significant (the p-values are also given as part of the output):

```
plot1 <- spplot(gwr1$SDF, "pctunemp", col.regions=colorRampPalette
   (c("red","orange","yellow"))(100))
```

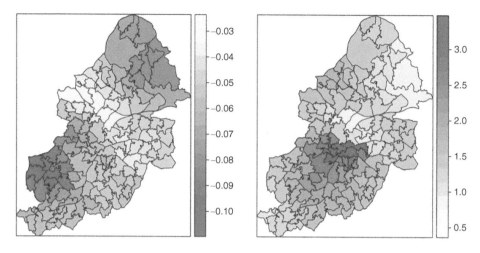

FIGURE 12.4   GWR estimates of the effects of (left) the unemployment rate on the log of the average house price and (right) the proportion of properties sold that were detached

```
plot2 <- spplot(gwr1$SDF, "detached", col.regions=colorRampPalette
  (c("yellow","orange","red"))(100))
grid.arrange(plot1, plot2, ncol=2)
```

This yields Figure 12.4.

Many other geographical operations are possible, including raster interpolation using the raster library or geometric operations such as buffering using the rgeos library. Suggested reading to learn about these packages is given at the end of this chapter.

12.3 A LITTLE MORE ABOUT THE WORKINGS OF R

Classes, types and coercion

Let us create two objects, each a vector containing ten elements. The first will be the numbers from 1 to 10, recorded as integers. The second will be the same sequence but now recorded as real numbers, also called 'floating point' numbers – those with a decimal place.

```
b <- 1:10
b
[1]  1  2  3  4  5  6  7  8  9 10
c <- seq(from=1.0, to=10.0, by=1.0)
```

```
c
[1]  1  2  3  4  5  6  7  8  9  10
```

Note that in the second case we could just type

```
c <- seq(1.0, 10.0, 1.0)
c
[1]  1  2  3  4  5  6  7  8  9  10
```

This works because if we don't explicitly define the arguments then R will assume we are giving them in their default order, which in this case is the argument from, the argument to, and the argument by, in that order. Type ?seq and look under usage for this to make a little more sense.

In any case, the two objects, b and c, appear the same on screen but one is an object of class integer whereas the other is an object of class numeric and of type double (double precision in the memory space):

```
class(b)
[1] "integer"
class(c)
[1] "numeric"
typeof(c)
[1] "double"
```

Often it possible to coerce an object from one class and type to another:

```
b <- 1:10
class(b)
[1] "integer"
b <- as.double(b)
class(b)
[1] "numeric"
typeof(b)
[1] "double"
class(c)
[1] "numeric"
c <- as.integer(c)
class(c)
[1] "integer"
```

```
c
[1]  1  2  3  4  5  6  7  8  9 10
c <- as.character(c)
class(c)
[1] "character"
c
[1] "1" "2"  "3"  "4"  "5"  "6"  "7"  "8"  "9"  "10"
```

The examples above are trivial. However, it is important to understand that seemingly generic functions like summary() can produce outputs that are dependent upon the class type. For example,

```
class(b)
[1] "numeric"
summary(b)
   Min.   1st Qu.   Median   Mean   3rd Qu.   Max.
   1.00      3.25     5.50   5.50      7.75  10.00
class(c)
"character"
summary(c)
   Length        Class        Mode
       10    character   character
```

In the first instance, a six-number summary of the centre and spread of the data is given. That makes sense for numeric data, but not for character data. The second summary gives the length of the vector, its class type and its storage mode.

A more interesting example is provided if we consider the plot() command, using it first with a single data variable, secondly with two variables in a data table, and finally on a model of the relationship between those two variables.

The first variable is created by generating 100 observations drawn randomly from a normal distribution with mean of 100 and a standard deviation of 20:

```
var1 <- rnorm(n=100, mean=100, sd=20)
```

Being random, the data assigned to the variable will differ from user to user. Usually we would want this. However, it can be fixed to ensure everyone gets the same 'random' draw:

```
set.seed(1)
# The 1 is arbitrary.
var1 <- rnorm(n=100, mean=100, sd=20)
```

Always check the data:

```
class(var1)
[1] "numeric"
length(var1)
[1] 100
# The number of elements in the vector
summary(var1)
   Min. 1st Qu.  Median    Mean 3rd Qu.    Max.
  55.71   90.12  102.30  102.20  113.80  148.00
head(var1)
[1]  87.47092 103.67287 83.28743 131.90562 106.59016  83.59063
# The first few elements
tail(var1)
[1] 131.73667 111.16973  74.46816  88.53469  75.50775  90.53199
# The last few elements
```

They seem fine. Returning to the use of the plot() command, in this instance it simply plots the data in order of their position in the vector:

```
plot(var1)
```

   To demonstrate a different interpretation of the plot command, a second variable is created that is a function of the first but with some random error:

```
set.seed(101)
var2 <- 3 * var1 + 10 + rnorm(100, 0, 25)
head(var2)
[1] 264.2619 334.8301 242.9887 411.0758 337.5397 290.1211
```

Next, the two variables are gathered together in a data table, of class data frame, where each row is an observation and each column is a variable:

```
mydata <- data.frame(x = var1, y = var2)
class(mydata)
```

```
[1] "data.frame"
head(mydata)
          x         y
1  87.47092 264.2619
2 103.67287 334.8301
3  83.28743 242.9887
4 131.90562 411.0758
5 106.59016 337.5397
6  83.59063 290.1211
nrow(mydata)
[1] 100
# The number of rows in the data
ncol(mydata)
[1] 2
# The number of columns
```

Now plotting the data frame will produce a scatter plot:

```
plot(mydata)
```

If there had been more than two columns in the data table, or if they had not been arranged in *x*, *y* order, then the plot could be produced by referencing the columns directly. All the following are equivalent:

```
with(mydata, plot(x, y))
# Here the order is x, y
with(mydata, plot(y ~ x))
# Here it is y ~ x
# ~ is the tilde character on your keyboard
plot(mydata$x, mydata$y)
plot(mydata[,1], mydata[,2])
# Plot using the first and second columns
plot(mydata[,2] ~ mydata[,1])
```

The attach() command can also be used:

```
attach(mydata)
plot(x, y)
detach(mydata)
```

To fit the regression model between y and x, use

```
model1 <- lm(y ~ x, data=mydata)
# lm is short for linear model
class(model1)
[1] "lm"
```

The summary() function summarises the relationship between y and x:

```
summary(model1)
Call:
lm(formula = y ~ x, data = mydata)
Residuals:
    Min      1Q   Median      3Q      Max
-57.102 -16.274    0.484   15.188   47.290
Coefficients:
              Estimate Std.   Error   t value   Pr(>|t|)
(Intercept)    8.6462        13.6208    0.635     0.527
x              3.0042         0.1313   22.878    <2e-16 ***
---
Signif. codes: 0 '***' 0.001 '**' 0.01 '*' 0.05 '.' 0.1 ' ' 1
Residual standard error: 23.47 on 98 degrees of freedom
Multiple R-squared: 0.8423,      Adjusted R-squared: 0.8407
F-statistic: 523.4 on 1 and 98 DF, p-value: < 2.2e-16
```

Using the plot() function on the object of class lm has an effect that is different from the previous two cases. It produces a series of diagnostic plots to help check the assumptions of regression have been met:

```
plot(model1)
```

The first is a check for non-constant variance and outliers, the second for normality of the residuals, the third is similar to the first, and the fourth identifies both extreme residuals and leverage points. These four plots can be viewed together, changing the default graphical parameters to show the plots in a 2 by 2 array,

```
par(mfrow = c(2,2))
# Sets the graphical output to be 2 x 2
plot(model1)
```

Finally, we might like to go back to our scatter plot and add the regression line of best fit to it:

```
par(mfrow = c(1,1))
# Resets the window to a single graph
plot(mydata)
abline(model1)
# abline sounds a lot like add line so is easy to remember
```

Data frames

The preceding section introduced the data frame as a class of object containing a table of data where the variables are the columns of the data and the rows are the observations:

```
class(mydata)
summary(mydata)
```

Looking at the data summary, the object mydata contains two columns, labelled x and y. These column headers can also be revealed by using

```
names(mydata)
[1] "x" "y"
```

or with

```
colnames(mydata)
[1] "x" "y"
```

The row names appear to be the numbers from 1 to 100 (the number of rows in the data), although actually they are character data:

```
rownames(mydata)
  [1] "1"   "2"   "3"   "4"   "5"   "6"   "7"   "8" [etc.]
class(rownames(mydata))
[1] "character"
```

The column names can be changed either individually:

```
names(mydata)[1] <- "v1"
names(mydata)[2] <- "v2"
```

```
names(mydata)
[1] "v1" "v2"
```

or all at once:

```
names(mydata) <- c("x","y")
names(mydata)
[1] "x" "y"
```

... as can the row names:

```
rownames(mydata)[1] <- "0"
rownames(mydata)
  [1] "0"   "2"   "3"   "4"   "5"   "6"   "7"   "8"     [etc.]
rownames(mydata) = seq(from=0, by=1, length.out=nrow(mydata))
rownames(mydata)
  [1] "0"   "1"   "2"   "3"   "4"   "5"   "6"   "7"   "8"   [etc.]
```

The above can sometimes be useful when merging data tables with GIS shape-files because the first entry in an attribute table for a shapefile is usually given an ID of 0. Otherwise, it is usually easiest for the first row in a data table to be labelled 1, so let's put them back to how they were:

```
rownames(mydata) <- 1:nrow(mydata)
rownames(mydata)
[1] "1" "2" "3" "4" "5" "6" "7" "8"     [etc.]
```

Referencing rows and columns in a data frame

The square bracket notation can be used to index specific rows, columns or cells in the data frame. For example:

```
mydata[1,]
          x         y
1 87.47092 264.2619
# The first row of data
mydata[2,]
          x         y
2 103.6729 334.8301
```

```
# The second row of data
round(mydata[2,],2)
       x      y
2 103.67 334.83
# The second row, rounded to 2 decimal places
mydata[nrow(mydata),]
            x        y
100 90.53199 261.236
# The final row of the data
mydata[,1]
  [1]  87.47092 103.67287  83.28743 131.90562 [etc.]
# The first column of data
mydata[,2]
  [1] 264.2619 334.8301 242.9887 411.0758 337.5397 [etc.]
# The second column, which here is also …
mydata[,ncol(mydata)]
  [1] 264.2619 334.8301 242.9887 411.0758 337.5397 [etc.]
# … the final column of data
mydata[1,1]
[1] 87.47092
# The data in the first row of the first column
mydata[5,2]
[1] 337.5397
# The data in the fifth row of the second column
round(mydata[5,2],0)
[1] 338
```

Specific columns of data can also be referenced using the $ notation:

```
mydata$x
  [1] 87.47092 103.67287 83.28743 131.90562 106.59016 [etc.]
# Equivalent to mydata[,1] because the column name is x
mydata$y
  [1] 264.2619 334.8301 242.9887 411.0758 337.5397 290.1211 [etc.]
summary(mydata$x)
   Min. 1st Qu.  Median    Mean 3rd Qu.    Max.
```

```
 55.71    90.12   102.30   102.20   113.80   148.00
summary(mydata$y)
   Min. 1st Qu.  Median     Mean 3rd Qu.      Max.
  140.4    284.1   314.1    315.6    355.7    447.6
mean(mydata$x)
[1] 102.1777
median(mydata$y)
[1] 314.1226
sd(mydata$x)
[1] 17.96399
# Gives the standard deviation of x
boxplot(mydata$y)
boxplot(mydata$y, horizontal=T, main="Boxplot of variable y")
```

One way to avoid the use of the $ notation is to use the function with() instead:

```
with(mydata, var(x))
[1] 322.7048
# Gives the variance of x
with(mydata, plot(y, xlab="Observation number"))
```

However, even this is cumbersome, so the attach() function may be preferred.

Attaching a data frame

Sometimes the ways to access a specific part of a data table become tiresome and it is useful to reference the column or variable name directly. For example, instead of having to type mean(mydata[,1]), mean(mydata$x) or with(mydata, mean(x)) it would be easier just to refer to the variable of interest, x, as in mean(x).

To achieve this the attach() command is used. Compare, for example,

```
mean(x)
Error in mean(x) : object 'x' not found
```

which generates an error because there is not an object called x in the workspace (it is only a column name within the data frame mydata), with

```
attach(mydata)
mean(x)
[1] 102.1777
```

which works fine. If, by an earlier analogy, objects in R's workspace are like box files, then you have now opened one up, its contents are visible and they include the variable x. To detach the contents of the data frame use **detach()**:

```
detach(mydata)
```

It is sensible to use detach when the data frame is no longer being used or else confusion can arise when multiple data frames contain the same column names, as in the following example:

```
attach(mydata)
mean(x)
[1] 102.1777
# This is the mean of mydata$x
mydata2 = data.frame(x = 1:10, y=11:20)
head(mydata2)
   x  y
1 1 11
2 2 12
3 3 13
4 4 14
5 5 15
6 6 16
attach(mydata2)
The following object(s) are masked from 'mydata':
    x, y
mean(x)
[1] 5.5
# This is now the mean of mydata2$x
detach(mydata2)
mean(x)
[1] 102.1777
detach(mydata)
rm(mydata2)
```

Subsetting the data table and logical queries

Subsets of a data frame can be created by referencing specific rows within it. For example, imagine we want a table of only those observations that have a value above the mean of some variable:

```
attach(mydata)
subset <- which(x > mean(x))
class(subset)
[1] "integer"
subset
  [1]   2  4  5  7  8  9 11 12 15 18 19 20 21 22 25 30 31 33 [etc.]
mydata.sub <- mydata[subset,]
head(mydata.sub)
           x         y
2 103.6729 334.8301
4 131.9056 411.0758
5 106.5902 337.5397
7 109.7486 354.7155
8 114.7665 351.4811
9 111.5156 367.4726
```

Note how the row names of this subset have been inherited from the parent data frame.

A more direct approach is to define the subset as a logical vector that is either true or false dependent upon if a condition is met:

```
subset <- x > mean(x)
class(subset)
[1] "logical"
subset
  [1] FALSE  TRUE FALSE  TRUE  TRUE FALSE  TRUE  TRUE  TRUE [etc.]
mydata.sub <- mydata[subset,]
head(mydata.sub)
           x         y
2 103.6729 334.8301
4 131.9056 411.0758
5 106.5902 337.5397
```

```
7 109.7486 354.7155
8 114.7665 351.4811
9 111.5156 367.4726
```

A yet more succinct way is:

```
mydata.sub <- mydata[x > mean(x),]
# Selects those rows that meet the logical condition, and all columns
head(mydata.sub)
          x         y
2 103.6729 334.8301
4 131.9056 411.0758
5 106.5902 337.5397
7 109.7486 354.7155
8 114.7665 351.4811
9 111.5156 367.4726
```

In the same way, to select those rows where x is greater than or equal to the mean of x, and y is greater than or equal to the mean of y:

```
mydata.sub <- mydata[x >= mean(x) & y >= mean(y),]
# The symbol & is used for and
```

Or, those rows where x is less than the mean of x, or y is less than the mean of y:

```
mydata.sub <- mydata[x < mean(x) | y < mean(y),]
# The symbol | is used for or
```

Missing data

Missing data are given the value NA. For example,

```
mydata[1,1] <- NA
mydata[2,2] <- NA
head(mydata)
          x         y
1        NA 264.2619
2 103.67287       NA
```

```
3   83.28743 242.9887
4 131.90562 411.0758
5 106.59016 337.5397
6   83.59063 290.1211
```

R will, by default, report **NA** or an error when some calculations are tried with missing data:

```
mean(mydata$x)
[1] NA
quantile(mydata$y)
Error in quantile.default(mydata$y) :
  missing values and NaN's not allowed if 'na.rm' is FALSE
```

To overcome this, the default can be changed or the missing data removed. To ignore the missing data in the calculation:

```
mean(mydata$x, na.rm=T)
[1] 102.3263
quantile(mydata$y, na.rm=T)
      0%      25%      50%      75%     100%
140.4475 282.1064 313.7536 356.5862 447.6020
# Divides the data into quartiles
```

Alternatively, there are various ways to remove the missing data. For example,

```
subset <- !is.na(mydata$x)
```

creates a logical vector which is true where the data values of x are not missing (the ! in the expression means not):

```
head(subset)
[1] FALSE  TRUE  TRUE  TRUE  TRUE  TRUE
```

Using the subset,

```
x2 <- mydata$x[subset]
mean(x2)
[1] 102.3263
```

More succinctly,

```
with(mydata, mean(x[!is.na(x)]))
[1] 102.3263
```

Alternatively, a new data frame can be created without any missing data whereby any row with a missing value is omitted:

```
subset <- complete.cases(mydata)
head(subset)
[1] FALSE FALSE  TRUE   TRUE   TRUE   TRUE
mydata.complete <- mydata[subset,]
head(mydata.complete)
          x         y
3  83.28743 242.9887
4 131.90562 411.0758
5 106.59016 337.5397
6  83.59063 290.1211
7 109.74858 354.7155
8 114.76649 351.4811
```

Reading data from a file into a data frame

Imagine a file called schools.csv containing information about the location and attributes of some schools. A standard way to read a file into a data frame, with cases corresponding to lines and variables to fields in the file, is to use the read.table() command:

```
?read.table
```

In the case of schools.csv, it is comma delimited and has column headers. Looking through the arguments for read.table() the data could be read into R using

```
schools.data <- read.table("schools.csv", header=T, sep=",")
```

This will only work if the file is located in the working directory, else the location (path) of the file will need to be specified or the working directory changed. It may be more convenient to use file.choose():

```
schools.data <- read.table(file.choose(), header=T, sep=",")
```

Looking through the usage of `read.table()` in the R help page, a variant of the command is found where the defaults are for comma-delimited data. So, most simply, we could use:

```
schools.data <- read.csv(file.choose())
```

The `write.csv()` function can be used to output data from R.

Lists

A list is a little like a data frame but offers a more flexible way to gather objects of different classes together. For example,

```
mylist <- list(schools.data, model1, "a")
class(mylist)
[1] "list"
```

To find the number of components in a list, use `length()`:

```
length(mylist)
[1] 3
```

Here the first component is the data frame containing the schools data. The second component is the linear model created earlier. The third is the character "a". To reference a specific component, double square brackets are used:

```
 head(mylist[[1]], n=3)
     FSM    EAL    SEN white blk.car blk.afr indian pakistani [etc.]
1 0.659 0.583 0.031 0.217   0.032   0.222   0.002   0.020
2 0.391 0.424 0.001 0.350   0.087   0.126   0.003   0.012
3 0.708 0.943 0.038 0.048 0.000   0.239   0.000   0.004

summary(mylist[[2]])
Call:
lm(formula = y ~ x, data = mydata)
Residuals:
   Min      1Q  Median      3Q     Max
-57.102 -16.274   0.484  15.188  47.290
Coefficients:
            Estimate Std. Error t value Pr(>|t|)
```

```
(Intercept)     8.6462     13.6208   0.635     0.527
x               3.0042      0.1313  22.878     <2e-16 ***
---
Signif. codes: 0 '***' 0.001 '**' 0.01 '*' 0.05 '.' 0.1 ' ' 1
Residual standard error: 23.47 on 98 degrees of freedom
Multiple R-squared: 0.8423, Adjusted R-squared: 0.8407
F-statistic: 523.4 on 1 and 98 DF, p-value: < 2.2e-16
class(mylist[[3]])
[1] "character"
```

The double square brackets can be combined with single ones. For example,

```
mylist[[1]][1,]
     FSM    EAL   SEN white blk.car blk.afr indian pakistani [etc.]
1 0.659 0.583 0.031 0.217   0.032   0.222  0.002    0.020
```

is the first row of the schools data. The first cell of the same data is

```
mylist[[1]][1,1]
[1] 0.659
```

Writing a function

In brief, a function is written in R in the follow way:

```
> function.name <- function(list of arguments) {
+ function code
+ return(result)
+ }
```

A simple function to divide the product of two numbers by their sum could be:

```
> my.function <- function(x1, x2) {
+ result <- (x1 * x2) / (x1 + x2)
+ return(result)
+ }
```

Running the function

```
my.function(3, 7)
[1] 2.1
```

Factors

Another class of object that is worth knowing about are factors. Consider the following example:

```
x <- 1:14
gp <- c(1, 1, 1, 2, 2, 3, 1, 1, 4, 4, 4, 2, 1, 2)
mydata <- data.frame(x, gp)
head(mydata)
```

As they stand, the numbers in the gp column of the data table are just integers:

```
class(mydata$gp)
[1] "numeric"
```

However, they could be set to denote a group to which the observation belongs:

```
mydata$gp <- factor(mydata$gp)
class(mydata$gp)
[1] "factor"
levels(mydata$gp)
[1] "1" "2" "3" "4"
```

These are no longer numeric but categorical values, hence the following command generates an error:

```
mean(mydata$gp)
[1] NA
Warning message:
In mean.default(mydata$gp) :
   argument is not numeric or logical: returning NA
```

However, the following makes perfect sense and produces four box plots alongside each other, one for each category:

```
boxplot(x ~ gp, data=mydata)
```

Sometimes data are read into R as factors when they are intended to be numeric, creating mistakes. They can be converted back in a slightly long-winded way that tends to avoid the potential for error, as.numeric(as.character(mydata$gp)).

12.4 CONCLUSION

This chapter has provided a brief overview of the inner workings of R and also of some of its geographical data handling capabilities. With very few exceptions, the vast majority of the analyses and graphics for this book were produced using R and the libraries listed at the beginning of the chapter.

As R has grown in popularity, the number of how-to guides for using it has increased markedly. Perhaps the best all round guide is *An Introduction to R*, which is freely available at CRAN or by using the drop-down Help menus in the RGui (R Core Team, 2015). It is clear and succinct. Of the many other books available, one of my favourites is Maindonald and Braun, *Data Analysis and Graphics using R* (2007). I also find useful: Adler, *R in a Nutshell* (2010); Crawley, *Statistics: An Introduction using R* (2005, a shortened version of *The R Book* by the same author (2007)); Field et al., *Discovering Statistics Using R* (2012); and Faraway, *Linear Models with R* (2015).[6]

None of these books is focused on mapping or spatial analysis, however. For that, the authoritative guide making the links between geographical information science, geographical data analysis and R is Bivand et al., *Applied Spatial Data Analysis with R* (2013).

A better book for novices, although itself covering some advanced material is, Brunsdon and Comber, *An Introduction to R for Spatial Analysis & Mapping* (2015).

Also helpful are Ward and Gleditsch, *Spatial Regression Models* (2008); and Chun and Griffith, *Spatial Statistics and Geostatistics* (2013). Both of these include examples of R code, as does Lovelace and Dumont, *Spatial Microsimulation with R* (2016).

There is plenty of free material too. For example, Lovelace and Cheshire, 'Introduction to visualising spatial data in R' (2014); Bivand, *Creating Neighbours* [i.e. spatial weights matrices] (2015); Baddeley, 'Analysing spatial point patterns in R' (2010); and Anselin, *Spatial Regression Analysis in R* (2005). My own *An Introduction to Mapping and Spatial Modelling in R* (Harris, 2013), from which some of this chapter has been taken, is also available.

Mostly I find help by searching online. Most of my questions have been answered already, and it doesn't usually take long to find what I am looking for. A wide range of documentation is listed on the R website,[7] and another good place to look is Emmanuel Paradis's *R for Beginners* (2005).[8]

---

[6]A predecessor of this book is available online: https://cran.r-project.org/doc/contrib/Faraway-PRA.pdf

[7]https://www.r-project.org/other-docs.html

[8]https://cran.r-project.org/doc/contrib/Paradis-rdebuts_en.pdf

POSTSCRIPT

It's 7.15 in the evening and very windy outside. It's also a few days since I wrote the preface, so no longer my birthday. It is, however, my sister's. She is a geography graduate too.

Being the evening, it will soon be time to put my children to bed. However, before I do so, there's another task more pressing. It's time to put this book to bed too. If you started at the beginning and have read this far, then you deserve a rest as well.

At the University I help to teach a course called 'Convincing Stories? Numbers as Evidence in the Social Sciences'. The idea is to provide an introduction to quantitative social science, looking at how numbers are used and abused to create 'stories' in the media, public policy, and in social and scientific debate. The aim of the unit is to prepare students for the sorts of methods and techniques they will encounter in their disciplines by discussing and debating the ideas and concepts that are used to create evidence in an uncertain world, and upon which decisions are made. The unit encourages students to engage critically with research and debate in their subject areas, placing them in a better position to learn quantitative skills, to conduct research and to enhance their studies. That's what the blurb says anyway. Sometimes we succeed (see Milligan et al., 2014).

As part of that course, I ask the participants to read a research paper I co-wrote called 'The changing interaction of ethnic and socio-economic segregation in England and Wales, 1991–2011' (Harris et al., 2015). The students are asked to go through the paper and to pick out all the different statistical techniques it contains and the ways they are used to tell a story. They include graphs, averages, correlations, index values and indices of segregation. I invite the students to pick holes in the paper. I pretend to have thick skin. I also ask the students to self-rate the techniques from easy to hard. Unsurprisingly, what I consider easy some regard as difficult. That is not a criticism of the students in any way. What I still grapple with, other researchers find pretty trivial. We are all at different stages in our learning.

I mention this for two reasons. First, because I am conscious that for a book about quantitative basics, some readers won't have found it basic at all, whereas others (probably long-serving professors) will mutter into their beards and wish I had gone deeper into certain topics. Both are right. It's simply a matter of perspective.

However, I hope I have provided an introduction to some core ideas, methods and concepts in quantitative geography and that, in doing so, it will have helped 'students to engage critically with research and debate in their subject areas'. Since that subject area is geography, I can express it more plainly: I hope I have helped geographers to engage with geography (or at least an important part of it).

The second reason is that I hope the engagement will indeed be critical but in the best sense of the word – informed, not merely negative, and with the aim of improving the research, not simply criticising it. Within the geographical literature right now there is a lot of innovation and creativity in the use of quantitative methods as we move from the small samples of the past to trying to make sense of complex surveys, administrative data, big data, volunteered data and other less traditional sources of information. Data are everywhere. Some of them are useful. The challenge is to find out which.

Dennis Prager is quoted as saying, 'our scientific age demands that we provide definitions, measurements, and statistics in order to be taken seriously. Yet most of the important things in life cannot be precisely defined or measured. Can we define or measure love, beauty, friendship, or decency, for example?' It's not often that I agree with a politically conservative radio talk show host, but in this instance I do. Quantitative geography is not the only thing that matters in geography, let alone in life. In the greater scheme of things, it may not matter that much at all. But it remains important precisely because definitions, measurements and statistics are taken seriously. Sometimes it is important to debunk that authority and show where the evidence is lacking. Sometimes it is important to create the evidence to better understand the world in which we live or to challenge what takes place within it – to provide evidence about the process that generate spatial inequalities and injustices, for example, or their consequences on things that matter like educational outcomes or life expectancy.

It's easy to find quotes about statistics. Apparently Jean Baudrillard said: 'Like dreams, statistics are a form of wish fulfilment.' Well, maybe, but then sometimes it's good to dream. It's now 11pm and I risk being flippant but I would appreciate that dreaming coming sooner rather than later.

REFERENCES

Abler, R., Adams, J.S. and Gould, P. (1972) *Spatial Organization: The Geographer's View of the World*. London: Prentice Hall.

Adler, J. (2010) *R in a Nutshell*. Sebastopol, CA: O'Reilly.

Adnan, M. and Longley, P. (2013) Tweets by different ethnic groups in Greater London. Featured Graphic. *Environment and Planning A*, 45(7), 1524–7.

Allen, R and Burgess, S. (2013) Evaluating the provision of school performance information for school choice. *Economics of Education Review*, 34, 175–90.

Allsop, R. (2013) *Guidance on Use of Speed Camera Transparency Data*. London: RAC Foundation.

Anselin, L. (1995) Local indicators of spatial association – LISA. *Geographical Analysis*, 27(2), 93–115.

Anselin, L. (2005) *Spatial Regression Analysis in R. A Workbook*. Center for Spatially Integrated Social Science, University of Illinois, Urbana-Champaign. https://geodacenter.asu.edu/system/files/rex1.pdf (accessed 1 December 2015).

Anselin, L. and Rey, S.J. (2014) *Modern Spatial Econometrics in Practice*. Chicago, IL: GeoDa Press.

Atkinson, A. (2015) *Inequality*. Cambridge, MA: Harvard University Press.

Baddeley, A. (2010) Analysing spatial point patterns in R. https://research.csiro.au/software/r-workshop-notes/ (accessed 1 December 2015).

Bartholomew, D.J. (2016) *Statistics Without Mathematics*. London: Sage.

Bell, A., Johnston, R. and Jones, K. (2014) Stylised fact or situated messiness? The diverse effects of increasing debt on national economic growth. *Journal of Economic Geography*, 15(2), 449–72.

Best, J. (2012) *Damned Lies and Statistics: Untangling Numbers from the Media, Politicians, and Activists* (updated edition). Berkeley/Los Angeles, CA: University of California Press.

Bivand, R. (2015) *Creating Neighbours*. https://cran.r-project.org/web/packages/spdep/vignettes/nb.pdf (accessed 1 December 2015).

Bivand, R.S., Pebesma, E. and Gómez-Rubio, V. (2013) *Applied Spatial Data Analysis with R* (2nd edition). New York: Springer.

Blastland, M. and Spiegelhalter, D. (2013). *The Norm Chronicles: Stories and Numbers About Danger*. London: Profile Books.

Bradley, I. and Meek, R. (2014) [1986] *Matrices and Society: Matrix Algebra and Its Applications in the Social Sciences*. Princeton, NJ: Princeton University Press.

British Academy (2013) *Stand Out and Be Counted: A Guide to Maximising Your Prospects*. London: British Academy. http://www.britac.ac.uk/policy/Stand_Out_and_Be_Counted.cfm

British Future (2013) *State of the Nation: Where Is Bittersweet Britain Heading?* London: British Future. http://www.britishfuture.org/wp-content/uploads/2013/01/State-of-the-Nation-2013.pdf

Brunsdon, C. (2015) Quantitative methods I. Reproducible research and quantitative geography. *Progress in Human Geography*, in press. doi:10.1177/0309132515599625

Brunsdon, C. and Comber, L. (2015) *An Introduction to R for Spatial Analysis & Mapping*. London: Sage.

Brunsdon, C., Fotheringham, A.S. and Charlton, M. (2002) Geographically weighted summary statistics – a framework for localised exploratory data analysis. *Computers, Environment and Urban Systems*, 26(6), 501–24.

Chang, W. (2013) *R Graphics Cookbook*. Boston, MA: O'Reilly Media.

Cheshire, J. (2012) Featured Graphic. Lives on the line: Mapping life expectancy along the London Tube network. *Environment and Planning A*, 44(7), 1525–8.

Cheshire, J. and Uberti, O. (2012) *London: The Information Capital: 100 Maps and Graphics that Will Change How You View the City*. London: Particular Books.

Chun, Y. and Griffith, D.A. (2013) *Spatial Statistics and Geostatistics*. London: Sage.

Cleveland, W.S. (1993) *Visualizing Data*. Summit, NJ: Hobart Press.

Crawley, M.J. (2005). *Statistics: An Introduction using R*. Chichester: Wiley.

Crawley, M.J. (2007). *The R Book*. Chichester: Wiley.

de Jong, P., Sprenger, C. and van Veen, F. (1984) On extreme values of Moran's *I* and Geary's *C*. *Geographical Analysis*, 16(1), 17–24.

Dilnot, A. and Blastland, M. (2008) *The Tiger That Isn't: Seeing Through a World of Numbers* (expanded edition). London: Profile Books.

Dong, G. and Harris, R. (2015) Spatial autoregressive models for geographically hierarchical data structures. *Geographical Analysis*, 47(2), 173–91.

Dong, G., Harris, R., Jones, K. and Yu, J. (2015) Multilevel modelling with spatial interaction effects with application to an emerging land market in Beijing, China. *PLoS One*, 10(6): e0130761.

Dorling, D. (2010) *Injustice: Why Social Inequality Persists*. Bristol: Policy Press.

Dorling, D. (2013) *The 32 Stops: The Central Line*. London: Penguin.

Dorling, D. (2014) *Inequality and the 1%*. London: Verso.

Dunn, R. (1989) Building regression models: The importance of graphics. *Journal of Geography in Higher Education*, 13(1), 15–30.

Dunning, T. (2012) *Natural Experiments in the Social Sciences: A Design-Based Approach*. Cambridge: Cambridge University Press.

Easterly, W. (2003) Can foreign aid buy growth? *Journal of Economic Perspectives*, 17(3), 23–48.

Faraway, J.J. (2015) *Linear Models with R* (2nd edition). Boca Raton, FL: CRC Press.

Few, S. (2012) *Show Me the Numbers: Designing Tables and Graphs to Enlighten* (2nd edition). Burlingame, CA: Analytics Press.

Field, A., Miles, J. and Field, Z. (2012) *Discovering Statistics Using R*. London: Sage.

Fioramonti, L. (2014) *How Numbers Rule the World: The Use and Abuse of Statistics in Global Politics*. London: Zed Books.

Foley, B. and Goldstein, H. (2012) *Measuring Success: League Tables in the Public Sector*. London: British Academy. http://www.britac.ac.uk/policy/Measuring-success.cfm (accessed 22 December 2015).

Fotheringham, A.S., Brunsdon, C. and Charlton, M. (2000) *Quantitative Geography: Perspectives on Spatial Analysis*. London: Sage.

Fotheringham, A.S., Brunsdon, C. and Charlton, M. (2002) *Geographically Weighted Regression: The Analysis of Spatially Varying Relationships*. Chichester: Wiley.

Galton, F. (1886) Regression towards mediocrity in hereditary stature. *Journal of the Anthropological Institute*, 15, 246–63.

Gelman, A. and Cortina, J. (eds) (2009) *A Quantitative Tour of the Social Sciences*. Cambridge: Cambridge University Press.

Goldacre, B. (2014) *I Think You'll Find It's a Bit More Complicated Than That*. London: Fourth Estate.

Gollini, I., Lu, B., Charlton, M., Brunsdon, C. and Harris, P. (2015). GWmodel: An R package for exploring spatial heterogeneity using geographically weighted models. *Journal of Statistical Software*, 63(17).

Gorard, S. (2014) The widespread abuse of statistics by researchers: What is the problem and what is the ethical way forward? *Psychology of Education Review*, 38(1), 3–10.

Gould, S.J. (1996) *The Mismeasure of Man* (2nd edition). New York: W.W. Norton.

Graham, M., Hale S. and Stephens, M. (2012) Featured Graphic. Digital divide: The geography of Internet access. *Environment and Planning A*, 44(5), 1009–10.

Gujarati, D. (2011) *Econometrics by Example*. Basingstoke: Palgrave Macmillan.

Hacking, I. (1990) *The Taming of Chance*. Cambridge: Cambridge University Press.

Hand, D.J. (2008) *Statistics: A Very Short Introduction*. Oxford: Oxford University Press.

Harmon, K. (2003) *You Are Here: Personal Geographies and Other Maps of the Imagination*. New York: Princeton Architectural Press.

Harris, R. (2013) *An Introduction to Mapping and Spatial Modelling in R*. http://www.researchgate.net/publication/258151270_An_Introduction_to_Mapping_and_Spatial_Modelling_in_R (accessed 1 December 2015).

Harris, R. (2014) Measuring changing ethnic separations in England: A spatial discontinuity approach. *Environment and Planning A*, 46(9), 2243–61.

Harris, R., Fitzpatrick, K., Souch, C., Brunsdon, C., Jarvis, C., Keylock, C., Orford, S., Singleton, A. and Tate, N. (2013) *Quantitative Methods in Geography: Making the Connections between Schools, Universities and Employers*. London: Royal Geographical Society (with IBG).

Harris, R. and Jarvis, C. (2011) *Statistics for Geography and Environmental Science*. London: Routledge.

Harris, R.J. and Johnston, R.J. (2008) Primary schools, markets and choice: Studying polarization and the core catchment areas of schools. *Applied Spatial Analysis and Policy*, 1(1), 59–84.

Harris, R., Johnston, R. and Burgess, S. (2016) Featured Graphic. Tangled spaghetti: Modelling the core catchment areas of London's secondary schools. *Environment and Planning A,* in press. doi: 10.1177/0308518X15603987

Harris, R., Johnston, R. and Manley, D. (2015) The changing interaction of ethnic and socio-economic segregation in England and Wales, 1991–2011. *Ethnicities*, in press. doi: 10.1177/1468796815595820

Harris, R., Sleight, P. and Webber, R. (2005) *Geodemographics, GIS and Neighbourhood Targeting*. Chichester: Wiley.

Hennig, B. (2013) Featured Graphic. The human planet. *Environment and Planning A*, 45(3), 489–91.

Hepple, L.W. (1998) Exact testing for spatial autocorrelation among regression residuals. *Environment and Planning A*, 30(1), 85–108.

Hern, A. (2013) The most misleading statistics of all, thanks to Simpson's Paradox. *New Statesman*, 2 May. http://www.newstatesman.com/2013/05/most-misleading-statistics-all-thanks-simpsons-paradox (accessed 1 December 2015).

Herrnstein, R.J. and Murray, C. (1996) *The Bell Curve: Intelligence and Class Structure in American Life*. New York: Simon & Schuster.

Holm, S. (1979) A simple sequentially rejective multiple test procedure. *Scandinavian Journal of Statistics*, 6(2), 65–70.

Huff, D. (1991) [1954] *How to Lie with Statistics*. London: Penguin.

Ioannidis, J.P.A. (2005) Why most published research findings are false. *PLoS Medicine*, 2(8), e124

Jacoby, R. and Glauberman, N. (eds) (1995) *The Bell Curve Debate: History, Documents, Opinions*. London: Times Books.

Johnston, R., Harris, R., Jones, K. and Manley, D. (2014a) A response to Gorard. *Psychology of Education Review*, 38(2), 5–8.

Johnston, R., Harris, R., Jones, K., Manley, D., Sabel, C. and Wang, W. (2014b) One step forward but two steps back to the proper appreciation of spatial science. *Dialogues in Human Geography*, 4(1), 59–69.

Johnston, R., Poulsen, M. and Forrest, J. (2015) Increasing diversity within increasing diversity: The changing ethnic composition of London's neighbourhoods, 2001–2011. *Population, Space and Place*, 21(1), 38–53.

Johnston, R. and Sidaway, J. (2015) *Geography and Geographers: Anglo-American Human Geography since 1945* (7th edition). London: Routledge.

Jones, K., Johnston, R., Manley, D., Owen, D. and Charlton, C. (2015). Ethnic residential segregation: A multilevel, multigroup, multiscale approach exemplified by London in 2011. *Demography*, 52(6), 1995–2019.

Kampstra, P. (n.d.) *Beanplot: A Boxplot Alternative for Visual Comparison of Distributions*. https://cran.r-project.org/web/packages/beanplot/vignettes/beanplot.pdf (accessed 1 December 2015).

Laszkiewicz, E., Dong, G. and Harris, R. (2014) The effects of omitting spatial effects and social dependence in the modelling of household expenditure for fruits and vegetables. *Comparative Economic Research*, 17(4), 155–72.

Leckie, G. and Goldstein, H. (2011) A note on 'The limitations of school league tables to inform school choice'. *Journal of the Royal Statistical Society, Series A*, 174(3), 833–6.

Leckie, G. and Goldstein, H. (2014) A multilevel modelling approach to measuring changing patterns of ethnic composition and segregation among London secondary schools, 2001–2010. *Journal of the Royal Statistical Society, Series A*, 178(2), 405–24.

LeSage, J., Pace, R.K., Campanella, R., Lam, N. and Liu, X. (2011) Do what the neighbours do: Reopening business after Hurricane Katrina. *Significance*, 8(4), 160–3.

Lloyd, C. (2014) *Exploring Spatial Scale in Geography*. Chichester: Wiley-Blackwell.

Longley, P.A., Goodchild, M.F., Maguire, D.J. and Rhind, D.W. (2015) *Geographic Information Science and Systems* (4th edition). Hoboken, NJ: Wiley.

Lovelace, R. and Cheshire, J. (2014) *Introduction to visualising spatial data in R*. National Centre for Research Methods Working Papers, 14(03). https://cran.r-project.org/doc/contrib/intro-spatial-rl.pdf

Lovelace, R. and Dumont, M. (2016) *Spatial Microsimulation with R*. London: Chapman & Hall/CRC.

Maindonald, J. and Braun, J. (2007) *Data Analysis and Graphics using R* (2nd edition). Cambridge: Cambridge University Press.

McCandless, D. (2012) *Information is Beautiful* (new edition). London: HarperCollins. Published in the USA as *The Visual Miscellaneum*.

McCandless, D. (2014) *Knowledge is Beautiful*. London: HarperCollins.

Meehl, P.E. (1978) Theoretical risks and tabular asterisks: Sir Karl, Sir Ronald, and the slow progress of soft psychology. *Journal of Consulting and Clinical Psychology*, 46(4), 806–34.

Milligan, L., Rose, J. and Harris, R. (2014) Convincing students? Quantitative junkies, avoiders and converts on a cross-disciplinary course using quantitative narratives. *ELiSS*, 6(2), 59–73.

Mitchell, R and Lee, D. (2014) Is there really a 'wrong side of the tracks' in urban areas and does it matter for spatial analysis? *Annals of the Association of American Geographers*, 10(3), 432–43.

Monmonier, M. (1996). *How to Lie with Maps* (2nd edition). Chicago, IL: University of Chicago Press.

Monmonier, M. (1999). *Air Apparent: How Meteorologists Learned to Map, Predict, and Dramatize Weather*. Chicago, IL: University of Chicago Press.

Monmonnier, M. (2001) *Bushmanders and Bullwinkles: How Politicians Manipulate Electronic Maps and Census Data to Win Elections*. Chicago, IL: University of Chicago Press.

Monmonier, M. (2010). *No Dig, No Fly, No Go: How Maps Restrict and Control*. Chicago, IL: University of Chicago Press.

Murphy, K.R. and Myors, B. (2003) *Statistical Power Analysis: A Simple and General Model for Traditional and Modern Hypothesis Tests* (2nd edition). Mahwah, NJ: Lawrence Erlbaum.

Ord, J.K. and Getis, A. (1995) Local spatial autocorrelation statistics: Distributional issues and an application. *Geographical Analysis*, 27(4), 286–306.

O'Sullivan, D. and Unwin, D. (2010) *Geographic Information Analysis* (2nd edition). New York: Wiley.

Owen, G., Harris, R. and Jones, K. (2015) Under examination: Multilevel models, geography and health research. *Progress in Human Geography*, in press. doi:10.1177/0309132515580814

Oxfam (2016) *An Economy for the 1%* (Oxfam Briefing Paper 210). http://policy-practice.oxfam.org.uk/publications/an-economy-for-the-1-how-privilege-and-power-in-the-economy-drive-extreme-inequ-592643 (accessed 2 February 2016).

Paloyo, A. (2014) When did we begin to spell "heteros*edasticity" correctly? *Philippine Review of Economics*, 2014, 51(1), 162–78.

Paradis, E. (2005) *R for Beginners*. https://cran.r-project.org/doc/contrib/Paradis-rdebuts_en.pdf

Piketty, T. (2014) *Capital in the Twenty-First Century*. Cambridge, MA: Harvard University Press.

Quality Assurance Agency (2014) *Subject Benchmark Statement: Geography*. Gloucester: QAA.

Parkhurst, D.F. (2006) *Introduction to Applied Mathematics for Environmental Science*. New York: Springer.

R Core Team (1999–2015) *An Introduction to R*. https://cran.r-project.org/ (accessed 1 December 2015).

Reinhart, C.M. and Rogoff, K.S. (2010) Growth in a time of debt. *American Economic Review*, 100(2), 573–8.

Robinson, W.S. (1950) Ecological correlations and the behavior of individuals. *American Sociological Review*, 15(3), 351–7.

Rogers, S. (2013) *Facts Are Sacred*. London: Guardian Faber Publishing.

Rogerson, P.A. (2014) *Statistical Methods for Geography* (4th edition). London: Sage.

Rozeboom, W.W. (1997) Good science is abductive, not hypothetico–deductive. In L.L. Harlow, S.A. Mulaik and J.H. Steiger (eds), *What If There Were No Significance Tests?* (pp. 335–92). Mahwah, NJ: Erlbaum.

Schelling, T. (1969) Models of segregation. *American Economic Review*, 59(2), 488–93.

Schelling, T. (1971) Dynamic models of segregation. *Journal of Mathematical Sociology*, 1(2), 143–86.

Siegel, D.A. and Moore, W.H. (2013) *A Mathematics Course for Political & Social Research*. Princeton, NJ: Princeton University Press.

Silver, N. (2013) *The Signal and the Noise: The Art and Science of Prediction*. London: Penguin.

Simpson, E. (1951) The interpretation of interaction in contingency tables. *Journal of the Royal Statistical Society, Series B*, 13(2), 238–41.

Singleton, A.D., Longley, P.A., Allen, R. and O'Brien, O. (2010) Estimating secondary school catchment areas and the spatial equity of access. *Computers, Environment and Urban Systems*, 35(3), 241–9.

Snow, J. (1855). *On the Mode of Communication of Cholera* (2nd edition). London: John Churchill.

Stiglitz, J. (2013) *The Price of Inequality*. London: Penguin.

Tan, S.T. (2012) *Applied Mathematics for the Managerial, Life, and Social Sciences* (international edition). Belmont, CA: Brooks/Cole.

Teele, D.L. (2014) *Field Experiments and Their Critics: Essays on the Uses and Abuses of Experimentation in the Social Sciences*. New Haven, CT: Yale University Press.

Thompson, B. (1996) AERA editorial policies regarding statistical significance testing: Three suggested reforms. *American Educational Research Association*, 25(2), 26–30.

Thompson, B. (2004) The 'significance' crisis in psychology and education. *Journal of Socio-Economics*, 33(5), 607–13.

Tiefelsdorf, M. (2002) The saddlepoint approximation of Moran's I's and local Moran's I_i's reference distributions and their numerical evaluation. *Geographical Analysis*, 34(3), 187–206.

Tobler, W.R. (1970) A computer simulation of urban growth in the Detroit region. *Economic Geography*, 46 (supplement), 234–40.

Tufte, E. (1990) *Envisioning Information*. Cheshire, CT: Graphics Press.

Tufte, E. (2001) *The Visual Display of Quantitative Information* (2nd edition). Cheshire, CT: Graphics Press.

Vigen, T. (2015) *Spurious Correlations*. New York: Hachette Books.

Ward, M.D. and Gleditsch, K.S. (2008) *Spatial Regression Models*. Thousand Oaks, CA: Sage.

Webster, M. and Sell, J. (2014) *Laboratory Experiments in the Social Sciences* (2nd edition). London: Academic Press.

Wilkinson, R. and Pickett, K. (2010) *The Spirit Level: Why Equality is Better for Everyone*. London: Penguin.

Wilson, R. (2003) *Four Colors Suffice: How the Map Problem Was Solved*. Princeton, NJ: Princeton University Press.

Wood, D. (1993) *The Power of Maps*. London: Routledge.

Wood, D. (2010) *Rethinking the Power of Maps*. New York: Guilford Press.

Ziliak, S.T. and McCloskey, D. (2008) *The Cult of Statistical Significance: How the Standard Error Costs Us Jobs, Justice, and Lives*. Ann Arbor, MI: University of Michigan Press.

Zook, M. and Graham, M. (2010). Featured Graphic. The virtual 'Bible belt'. *Environment and Planning A*, 42(4), 763–4.

INDEX

Page numbers in **bold** indicate tables and in *italic* indicate figures.